MW01002150

THE IDEOLOGICAL BRAIN

THE IDEOLOGICAL BRAIN

The Radical Science of Flexible Thinking

LEOR ZMIGROD

Henry Holt and Company
New York

Henry Holt and Company
Publishers since 1866
120 Broadway
New York, New York 10271
www.henryholt.com

Distributed in Canada by Raincoast Book Distribution Limited

Library of Congress Cataloging-in-Publication Data is available.

ISBN 9781250344595

Our books may be purchased in bulk for promotional, educational, or business use. Please contact your local bookseller or the Macmillan Corporate and Premium Sales Department at (800) 221-7945, extension 5442, or by e-mail at MacmillanSpecialMarkets@macmillan.com.

First U.S. Edition 2025

Designed by Gabriel Guma

Printed in the United States of America

10 9 8 7 6 5 4 3 2 1

Dedicated with love to my family

All day long, all through the night,
all affairs—yours, ours, theirs—
are political affairs.

Whether you like it or not,
your genes have a political past,
your skin, a political cast,
your eyes, a political slant.

Whatever you say reverberates,
whatever you don't say speaks for itself.
So either way you're talking politics.

—WISŁAWA SZYMBORSKA,
"CHILDREN OF OUR AGE"

CONTENTS

THE IDEOLOGICAL BRAIN

PROLOGUE

ACTION POTENTIAL

All we need is conviction. Convictions offer us certainty—or, at least, the *appearance* of certainty when we are in fact unsure. Convictions reveal our deepest passions—or, at least, give us things to be passionate about. Convictions bring us together with other people through a common and dedicated purpose, creating a loving community out of mere strangers. *How joyful!* If all these convictions coalesce into a worldview that is reasonably coherent, we can triumphantly declare that we have an ideology: a set of truths and moral principles that we live by and share with others. *It's easy!*

Conviction is all we need. Whether our ideological mission is ancient or new, religious or secular, conservative or reactionary, communicated digitally or in the flesh, it prepares us to distinguish right from wrong, good from evil, an ethical decision from one that is foolish or selfish. Under the guidance of wise authorities, we imagine heavenly utopias to build on earth and devise strategies to avert oncoming disasters and moral catastrophes. We begin to embrace new symbols, adopt a new fashion, fit into a new family, participate in beguilingly opaque rituals, feel the ecstasy of dissolving into a collective that accepts our membership with jubilation.

Our brain—an organ that constantly seeks to understand the world around us, to feel belonging with others—is delighted by our newfound ideological possessions.

What could possibly go wrong?

I sat in a darkened university laboratory—a tiny room enclosed by black walls. The dimmed lab was usually occupied by neuroscientists of sleep who laid participants down on makeshift beds and measured the electrical activity of their brains at rest, drifting off to dream. Yet I was interested in the opposite of slumber: I was there to detect the neural signature of choice, of free will. Over a long summer, I stuck electrodes on the scalps of participants and watched their brain waves dance on the monitor—fidgety lines climbing and collapsing, rendering invisible processes visible. In these experiments, I studied what brains look like when they obey commands as opposed to when they form free and spontaneous decisions. With the techniques of neuroscience, I learned that obedient actions evoked neural activity patterns that were markedly different from free choices.

At the time, I was a psychology student at Cambridge interested in sensory perception and the neuroscience of free will. I was passionate about the potential of neuroscience to answer fundamental questions about human consciousness—what it feels like to be aware or unaware of a sensation, how unconscious impressions are formed—and so I volunteered as a research assistant for my professors' research projects on weekends and holidays.

I spent the summer's bright afternoons in a little windowless experiment room, gluing sensors to skin with sticky jelly, installing and reinstalling a lattice of metallic discs and wires on participants' heads. In the evenings I would analyze the results, zooming into one of the smallest units of neuroscience: the electrical potential that precedes every movement, voluntary or coerced. Under the glow of pixelated neural impulses, I searched for the subterranean, unconscious markers of human freedom.

But this was 2015, and outside the sunless laboratory room new forms of fundamentalism were on the rise. When I heard news about young British girls being drawn to go to Syria to join ISIS, a question tugged at me: Why were these particular girls lured into extremism? Many commentators chalked it up to demographic factors and the dangers of the internet: the foolishness of youth, a lack of liberal education, the perils of financial or cultural precarity. But these felt like insufficient explanations. Many people endure socioeconomic and technological risks. Demography is not destiny. So why did these particular girls join an ideological war that led to their expulsion from home and the erasure of their freedoms? *Why them and not others?* Perhaps demographics and folk psychology could not capture the full story—maybe there was something about their brains that made these young people vulnerable.

I was curious to see whether I could connect the cognitive and neuroscientific methods I had grown to love and apply them to the domain of politics—to questions about ideologies. Could a person's susceptibility to extreme worldviews be rooted in the idiosyncrasies of their cognition and biology? Was it possible that human consciousness could be fundamentally altered by adherence to dogmatic ideologies?

Beginning my experiments in the tumultuous months following the United Kingdom's Brexit referendum and just before the 2016 US presidential election, I was among the first wave of scientists to use cognitive and neuroscientific methods to investigate the origins and consequences of ideological thinking. I recruited participants online who spanned all walks of life and held views that traversed the traditional to the ultraprogressive—from radical activists writing on right-wing platforms to German adolescents living in a reunified Berlin and elderly pensioners in remote British villages. I harnessed novel methods that would allow thousands of participants to complete the experiments from the comfort of their own homes, and I collaborated with international colleagues to collect brain scans and genetic samples of selected participants in university laboratories.

It was a rare choice to adopt methods that used cognitive assessments and brain-scanning technology to investigate ideologies. Only a handful of international research teams were interested in bringing the biological and political sciences together. It was a high-risk, high-reward research strategy. And it paid off.

With modern scientific techniques we are now able to ask how deeply into the architecture of the human brain ideological systems can penetrate—how far into the mind and body indoctrination really goes. We have discovered that the brain schooled by ideology is a brain worth exploring. A close study uncovers what ideologies can do to our bodies and how harsh moralities slip into the deepest recesses of human consciousness. It illuminates who has the potential for extremism; why some brains are particularly vulnerable while others are more flexible and resilient.

The young British adolescents who were radicalized in their bedrooms, at sleepovers, on smartphones, were extraordinary cases of ordinary processes—processes that every brain is susceptible to, and that some brains are more receptive to than others. The clues for the differences in risk between us lie in our cells, our bodies, our personal narratives.

Although a dogmatic environment can produce habits and compulsions that appear to the outside observer to be passive and automatic—almost without thought—when we investigate the ideological brain we see that there are sophisticated and dynamic processes happening inside. Neurons buzz, fire in synchrony, and shoot action potentials with every dutiful step. The origins of our ideological convictions emerge from within our bodies and the outcomes of our ideological beliefs can be felt and seen within our bodies too.

This book weaves together neuroscience and politics and philosophy to challenge our understanding of what it means to exist as human beings awash in dogmas, trying to stay afloat among raging orthodoxies. It can be read with different ideologies in mind—nationalistic movements, religious ideologies, racist worldviews, conspiratorial cults, "far right" and "far left" and political ideologies that are maybe not far away enough.

Although we will engage with the science of beliefs and the results of laboratory experiments, the critique of ideologies is not an exercise of pure reason. It has practical implications. It must reckon with people's emotional investments, such as their love of tradition and history; of groups and collectivities; of principles and guiding moral laws and categories; of faith and dedication, and of the people we hold in memory when we hear the word "ideology." Many kinds of love are on the line. The stakes are immeasurably high.

This book conveys a new and radical science that urges us to reimagine our ideologies and the risks involved in adopting rigid ways of thinking. It reveals that our politics are not superficial: our politics can become cellular. We will zoom into the ideological brain with a scientist's microscope, a philosopher's concern, a humanist's hope, and an active citizen's empathy and imagination—hoping that in the contrasts of openness and hate, revision and tradition, evidence and imposed fates, we will uncover what the free, authentic, and tolerant brain looks like too.

Part I

ICONS

1

IDEOLOGICAL POSSESSION

We say people "have" an ideology, as though it were a suitcase or a banana. Like objects we can hold, cherish, or discard, ideologies are imagined as being external to us. Sometimes we exchange an old ideology in favor of a newer, shinier one. Other times *we* are the evangelists trying to push an ideology into the palms of the unconvinced. *Take it!*

We barter and trade these ideological possessions, boasting about the values of our latest acquisitions. Yet maybe we are mistaken in thinking that ideologies are goods we hold, baggage we carry, that ideologies somehow exist outside of us.

We possess beliefs, yes, but we can also become possessed by them. With powerful measurement tools, it is now possible to see the consequences of ideological rigidity all the way down to human perception, cognition, physiology, and even neural processes. Our bodies are not impervious to the ideologies that surround us: what we believe is reflected in our biology.

Unlike the impressions left in sand, ideological imprints are difficult to erase. Our most recent scientific discoveries illustrate that human brains soak up ideological convictions with vigor and thirst. After all, our brains are magnificent organs that learn from their environments easily. Dangerously quickly. So when submerged in dogmatic systems, our bodies willingly absorb such rigidities. Repeating rules and rituals, rules and rituals, has stultifying effects on our minds. With every reiteration and rote performance, the neural pathways underpinning our habits strengthen, whereas alternative mental associations—more original yet less frequently rehearsed—tend to decay. While many of us intuitively know that ideologies dictate our social behaviors and moral sympathies, it is less well known that the repetition of ideological rules and rituals cascades into our cells.

Immersion in rigid and authoritarian structures is not simply a social or political problem. It is a profoundly personal problem for each of us. Ideologies can endanger the health of our minds and our capacities for authenticity. Our bodies learn to embody ideological convictions in deep and troubling ways. Unless we understand what ideologies are and what they do to us, different extremisms will emerge and mutate, advancing unhindered into our open and tolerant societies. And until we uncover how the brain metamorphoses under the grip of ideological doctrines, genuine freedom will continue to elude us.

Ideologies are sold to us as timeless and fixed, but they are in fact highly fluid and mobile. These clusters of ideas are perpetually changing, taking on novel guises in every generation. Ideological worldviews can switch sides and policy preferences. Parties of tradition campaign for radical reform while progressive movements hesitate to innovate. Guns are lifted in the name of life. Slogans for peace are used to camouflage regressive violence. Terrorism can hijack the fight for freedom and demands for freedom can appear terrorizing.

Battles over ideologies resemble language games. Words are thrown about, rhetorical devices are hurled at opponents and narrowly sidestepped. *Reactionary, revolutionary, conservative, progressive, conspiracist, supremacist, racist, radical, bigot.* We rarely know what these labels mean or to whom they rightly refer. George Orwell observed that "political language . . . is designed to make lies sound truthful and murder respectable, and to give an appearance of solidity to pure wind." We assign people and ideas into neat categories in the pursuit of clarity and identity. *Our neighbor is a fanatic! Our teenager is a fool!* Such taxonomies delight and shock us. Yet these linguistic buckets mask the realities of ideologies as they are lived—messily, hypocritically, proudly, self-destructively—with loss, joy, humor, regret, fear, reversals, retractions, ruminations, intimacy, and grief—with tears and lamentations, beaming smiles, and confused sideways glances.

Despite the complexities and contradictions, there are commonalities in how ideologies are practiced and preached, regardless of their aims or their claims. Whether nationalist or racist or religious, there are parallels in how all ideologies infiltrate human minds. These resemblances are not coincidences; they are inherent to the structure of ideological thinking. As the political thinker Eric Hoffer observed in *The True Believer*, "there is a certain uniformity in all types of dedication, of faith, of pursuit of power, of unity and of self-sacrifice." While ideologies might be dressed in different colors or costumes or flags, there is evidence that across different ideological groups the mechanisms of ideological coercion are largely the same.

To detect the psychological similarities across ideologies, we need a sense of what an ideology is and what it is not. In its simplest formulation, an ideology is a kind of narrative. A compelling story about the world. Yet not all stories are ideologies, and not all forms of collective storytelling are rigid and oppressive. There is a difference between culture and ideology. Ideologies offer absolutist descriptions of the world and accompanying prescriptions for how we ought to think, act, and interact with others. Ideologies legislate what is permissible and what is forbidden. Unlike culture—which can celebrate eccentricities and reinterpretations—in ideology, nonconformity is intolerable and total alignment is essential. When deviation from the rules leads to severe punishment and ostracism, we have moved away from culture and into ideology.

From fascism and communism to eco-activism and spiritual evangelism, ideological groups offer absolute and utopian answers to societal troubles, strict rules for behavior, and an ingroup mentality through dedicated practices and symbols. These features exist across the spectrum of ideological persuasions. Such characteristics can emerge even when the ideology is guided by the sincerest intentions and noblest ideals—even if it claims to protect human dignity or flourishing.

Typically, ideologies are imagined as big visions. Grand and atmospheric. Intangible and out of our personal control. Few of us can outline the precise tenets of pompously uppercased Conservatism, Liberalism, Fascism, Communism, Capitalism, Racism, Sexism, Theism, or Populism, with all their myriad meanings and interpretations. As though from the heavens, these *-isms* describe the contours of life and prescribe human action, instructing us about the cosmos and how we ought to relate to others within it. For believers, the utopian destiny of an ideology seems carved from the clouds of eternity. A looming force soaring above our heads, meant to be venerated and revered.

The image of ideologies as celestial and static has always troubled me. Ideologies coexist among us, within us, on earth. Not in the skies of history or the towers of political elites. There is no transcendent plane on which they live; no altitudes from which attitudes descend fully formed and holy. Ideologies inhabit individuals. Individual minds convert social doctrines into ideological thinking, a style of thinking that is governed by strict mental rules and carefully regimented mental leaps.

While most definitions perceive ideologies as historical currents and sociological movements, I am interested in examining ideologies as psychological phenomena instead. This psychological lens allows us to ask what an ideology does to its believers and whom it most easily attracts. By spotlighting the processes happening within individual brains, we can probe when an ideology constrains its followers' mental lives and whether it can ever liberate them.

Even if an ideology seems righteous, ethical, vital, urgent, or beautiful, I believe it should be examined closely. We can study an ideology's structure, its genesis and effects, what it alters in adherents' minds. We can scrutinize what, in a mind, an ideology fractures or silences; which biological and mental processes an ideology distorts. Does the ideology impose a tight grip on believers' brains? Or does it let them wonder and wander freely?

Every worldview can be practiced extremely and dogmatically. Every kind of cultural narrative used to explain the world can tip into a totalizing ideology. As a result, inquiring into *what* an ideology urges us to think is insufficient; we need to analyze *how* it makes us think too. When an ideology demands rigid and ritualistic thinking, it demands that we bias our vision, twist our gnawing doubts into silence, surrender our subjectivities and creative possibilities. When an ideology demands rigid and ritualistic thinking, it demands that we become someone else. Someone less singular and unique, less curious, less free.

Traditionally, we judge an ideology based on its merits, faults, and leaps of logic. *Bias, bias, bias,* we announce. Mining our opponents' belief systems, we unearth contradictions and hypocrisies. Different layers of naivety or callousness or ignorance that deserve scorn or ridicule. We criticize the viewpoints of our adversaries for their legal or economic assumptions, their social ills, or historical similarities to older worldviews.

I hope to show that we can challenge ideologies on different terms—on the terms of the single individual. The individual brain. I believe that we can judge an ideology based on what believing in it does to human bodies and brains; on whether being a passionate believer narrows our movements, lassoes our flexibility, restricts our responses, or triggers us to commit violence. If we have less scope for plasticity and change and less direct access to our sensations, we are at risk of dehumanizing ourselves and others. We become less sensitive, less elastic, less authentic. If we see reality through an ideological lens, we end up avoiding the richness of existence in favor of a more reduced, stereotyped experience. By studying the ideological brain with neuroimaging devices and cognitive tests, we can illuminate previously invisible forms of domination. With the tools of science, we can develop new ways to critique ideologies.

Perhaps some ideologies will pass our critical tests. Many will not. We might accidentally become suspicious of our most treasured ideological possessions. A science of ideology can inspire us to question our idols, our icons, our metaphors, our imagined utopias. It can stimulate careful analysis and honest self-reflection. It can even become the basis for personal or social action. Examining the neurocognitive origins and consequences of our beliefs—where they come from and how they transform our bodies—will offer clues regarding the kind of belief systems we might wish to keep and which ones we might be persuaded to let go.

Believing passionately in a rigid doctrine is a process that spills into our neurons, flowing into our bodies. Ideologies are not mere envelopes for our lives; they *enter* our skins, our skulls, our nerve cells. Totalizing ideologies shape the brain as a whole, not simply the brain when it is confronted with political propositions or debates. Science is beginning to reveal that the profound reverberations of ideologies can be observed in the brain even when we are not engaging with politics at all. Since our brains learn to embody indoctrination in deep and insidious ways, the social rituals we learn to enact can become the biological realities of our minds and bodies. There is therefore a danger that when an individual is immersed in rigid ideologies, it is not only their political opinions and moral tastes that are being sculpted—their entire brain is being sculpted too.

2

AN EXPERIMENT

I invite you to sit down on that gray chair—yes, the one at the desk—and make yourself comfortable. I point to the monitor in front of you and say this is where the experiment will happen. Soon, when I leave the room, you will see instructions pop up on the screen. Please read the instructions carefully. If you have any questions, ring the bell or simply raise your arm and wave, and I'll be right with you. The whole experiment will last a few minutes. As per ethics regulations and protocols, you are free to stop the experiment at any time, but please try not to.

Does that sound good?

You nod tentatively.

Perfect. Good luck!

Please press ENTER when you are ready.

You ENTER.

Hello! Welcome to the experiment. Today you will play a series of brain games and problem-solving challenges. For the first game, you will be presented with a deck of cards. Each card will be painted with a number of geometric objects of a specific color and shape. For instance, you may encounter a card with three red circles or a card decorated with a single blue triangle.

The game is a "card-sorting task." A card will appear at the bottom of your screen. Imagine it is painted with four orange squares. You need to decide how to match it to one of four cards already at the top of the screen.

✓♪ You will hear a happy jingle when you choose the CORRECT match.

✕♪ You will hear an angry beep when you choose the INCORRECT match.

Be as accurate and as quick as you can.

Please press ENTER if you understand the instructions.

Press ESCAPE if you wish to read the instructions again.

You ENTER.

Your first card has three green stars.

You try to match it with the card at the top of the screen decorated with the two blue stars. Maybe stars should go together with other stars.

BEEP!

You sigh. You try again. Maybe your three green stars should be paired with the card containing four green circles? Green-on-green?

Drag, press, release, and . . . happy jingles! You are right!

You shrug proudly to yourself.

Green-on-green. Easy.

Next card in your deck: one red triangle.

You follow the rule: pair color with color. You place red on red and . . . jackpot! Jingles again.

You like this rule. You apply it on the next round and the next. Green-on-green, red-on-red, orange-on-orange, blue-on-blue.

The habit is oddly fulfilling. Sliding cards to their rightful grouping, you barely need to think. It becomes wonderfully automatic. Color. Color. Color. Moments ago you were new to this world and now you are its master.

All you see are primary colors. You forget the other features of the cards. Nothing else matters.

You click, you match, you hear the saccharine jingle.

Color—jingle. Color—jingle. Pavlov would be proud.

You develop a ritual and it's glorious. It offers you control.

After five, or ten, or fifteen rounds—repetition blurs the boundaries of time—the next card in your deck has two blue squares. You know what to do. Go for the blue card at the top of the screen. Caressing the mouse, you point the cursor and shoot, waiting for the chime, anticipating the dopamine rush.

BEEP!

An angry, unexpected noise is emitted from the speakers.

You feel betrayed. You forgot the game world was capable of such an offensive sound. It's insulting.

Maybe it's just a glitch.

You select the blue card again. It's second nature to you now, blue on blue.

BEEP!

How can this be? The game world's inconsistency is like an astonishing infidelity. It makes you want to get up and leave the experiment room.

But you are an addict now. The jingle gave you the feeling (*the illusion?*) of control, of self-possession. It signaled your cleverness.

And now abandoned, you have no anchor, no habit, no ritual to turn to. *In this suddenly desolate virtual land, what will become of you?*

In a mad rush you drag the two-blue-squared card toward the three-orange-circled card—there is nothing unifying these cards, not number or color or shape, but you don't care, you are annoyed. BEEP! The noise barely dissipates before you are lugging the card again, this time toward the four-green-starred card. BEEP! Outraged at this rebellion, you move the mouse in fast, frenzied motions. You thought you had a deal with the game world. An understanding. Common ground on which to play. The rules are not supposed to change halfway through the game. You haul the card to the last unexplored option, swearing to yourself that if this isn't a match, if the jingle doesn't return, you will storm out of here in protest, you will wave your arm in the air to call the experimenter back in the room and demand answers, you will—jingle! It worked! You strain your eyes to see why the longed-for tune returned, what the matching card was. It was two red triangles. Two. Two! Hah! The number of shapes on the card was the same as on the card you held. Hallelujah! Maybe order will return once more. Or maybe this iteration of the task was just a bug. A mere hiccup.

Next time a card pops up on the screen, should you obey the old tradition, follow the color code, or try this new pattern, count the numbers and sort anew? Should you stick to your guns—ignore the anomaly—or should you change, explore, adjust, adapt, revise, and realize that—

This is where I step out of the experiment and tell you that your natural reaction to the change can tell me almost everything about you. Your spontaneous response to the fact that the old rule stopped working and you needed to discover a new one to survive is a kind of inadvertent confession. In this simple game of stars and circles, you have accidentally and inevitably exposed your innermost beliefs.

Why? Because there are two of you. There is the participant who notices the change in the rule governing the game and responds by changing in line with the new demands of the task. This version of you is the adaptable, cognitively flexible individual. When the world changes, you may feel surprise but you have no fear. You change with the times, with the demands of the environment. You are not strongly rule-bound. You are happy to slip between habits. In fact, you don't mind having no habit at all. You easily switch between modes of thinking; you are fluid; elastic; you adapt.

However, there is another you. In this version of you, you hate the change. You notice the fact that the old rule no longer works, and you refuse to believe it. You will try again and again to repeat the first rule, but it will be in vain. In fact, you will be punished every time you repeat the original habit. The unnerving BEEP will hit you like a slap in the face. But you won't move, won't dodge the blow. You will remain immobile, hanging on tightly to the false belief that somehow the wrathful beep will dissipate and be replaced by a jolly melody. The false and nostalgic belief that the environment around you will magically return and so you don't need to change. You persevere even when it would be faster to sever ties with the past and move on. This is the cognitively rigid version of you.

Which of these copies of you are you? The flexible or the rigid? The adaptable or the stubbornly unmoving?

Maybe you are neither the first nor the second. You could be somewhere in between: sometimes adaptable, sometimes rigid. Maybe your flexibility depends on circumstance. At ease, you are fluid, adjusting calmly to novelty or surprise. Yet in moments of stress, your movements narrow, your thoughts harden. Anxiety solidifies you, rendering you stiff.

What I, the experimenter, the scientist, have discovered is that how you perform in this game can give me clues about your whole approach to life. Your level of rigidity in this neuropsychological test foreshadows the rigidity with which you believe in ideologies in the social and political world. Your perceptual reflexes are linked to your ideological reflexes.

In fact, your brain comes to mirror your politics and prejudices in strange, profound, and astonishing ways—ways that challenge how we understand the tensions between nature and nurture, risk and resilience, freedom and fate. If our ideological beliefs are related to our cognitive and neural patterns of responding, then we must face new questions about how our bodies become politicized and in what ways we are capable of resisting, changing, and exercising personal agency.

When my colleagues and I invited thousands of people to complete cognitive tests of mental flexibility such as this game, called the Wisconsin Card Sorting Test, we found that the people who are the most behaviorally adaptable on neuropsychological tasks are the same people who—in the realm of ideologies—are most open-minded, most accepting of plurality and difference. The people with the most flexible minds are the people who acknowledge that the intellectual realm can be separated from the personal realm. They do not viscerally hate their interlocuters—they may hate their opinions but they do not project that hatred onto the persons voicing them. In contrast, the most cognitively rigid individuals, those who struggle to change when rules change, tend to hold the most dogmatic attitudes. They hate disagreement and are unwilling to shift their beliefs when credible counterevidence is presented.

Cognitive rigidity translates into ideological rigidity.

This may seem obvious to some: a rigid person is a rigid person. But in fact these patterns are not obvious. When neuroscientists talk about cognition and perception, we are talking about information processing that deals with simple stimuli, with basic sensory information in neutral contexts. Cognitive tasks are composed of uncomplicated elements—colored shapes and moving black dots—displayed on spare, undecorated screens. Through these tasks, we are not assessing how you deal with emotionally evocative or triggering information—information that genuinely scares you or makes you feel a sour pinch of disgust. We are not studying tasks that are too cognitively demanding or complex—ones that would exasperate you needlessly. When neuroscientists measure cognition and perception, we glean individual differences in how a brain forms decisions, learns from the environment, and responds to challenges or contradictions at the most foundational level.

These individual differences are implicit; we have little conscious access to them or control over their expression. A cognitively rigid person may insist that they are spectacularly flexible, and an adaptable thinker may believe that they lack mental malleability. It is astonishing how rarely we know ourselves.

As a result, the link between mental inflexibility and ideological rigidity reveals a critical insight about how our brains work and how ideologies penetrate human brains. It suggests that our characteristic rigidity, rigidity that is evident when we deal with any information—even orange stars and blue circles—can propagate up to higher-level rigidities that emerge in our ideological choices and actions.

Even when we are not thinking explicitly about politics, the reverberations of our ideological convictions can be felt and measured. Ideological imprints on the brain can be observed when our minds are left to roam and drift, when we imagine and invent, when we observe and interpret even the most neutral of situations. The ideological brain's rigidities and idiosyncrasies manifest where we least expect them, in our most private sensations and physiological responses, beneath the surface of our public convictions and conscious feelings. The dangers of dogmatic ideologies are therefore not just political—the consequences are neural, individual, and existential.

3

METAPHORS WE BELIEVE BY

When we imagine a body spellbound by an ideology—a mind enamored by ritual and rigidity—we encounter a figure without a clear face. The metaphors we use to describe a brain invaded by an ideology sustain this mystery and fuzzy kind of vacancy. There are metaphors of a mind being emptied. A mind being hypnotized. A mind deluded or sedated. A mind infantilized, sent back to childhood. A mind stupefied, dizzy with falsehoods.

Some of these analogies rely on imagery of overwhelming fullness: a mind saturated with wrong ideas whizzing around. The ideologue overflows with delusions and feelings; full of hate, full of fear, full of disgust, flooded with frustrated-love and power-lust. It is only *lies, lies, lies* that swirl behind their eyes. Simplified slogans bounce around the skull with lavish excess. *Return to the past! March to the future! Everywhere lurk citizens of nowhere . . .*

In these metaphors, the believer is full of bad things, and too much of a bad thing is never wonderful. Bloated with illusions, the ideologue loses their poise and self-possession, and simply becomes possessed.

In other analogies, the convert's mind is empty and hollow. *Knock, knock.* There is nothing there. No thoughts. No mind. A false consciousness. Just a shell. Ideologies suck all substance out, and only a void remains. A zealot is a zombie, obeying orders mechanically, blankly, almost unconsciously.

This sense of deletion is already in our language: to be brainwashed is to be washed of a brain, cleansed of a mind. The capacity for independent thought is sanitized and scrubbed away.

Although these metaphors can be convenient, and sometimes comforting—*unlike those of other people, our own brains are not brimming with lies or barren with nothingness*—such metaphors can deceive. Ideological thinking is not merely an excess of myth or a scarcity of thought. Our minds are not passive vessels to be filled, drained, and refilled. What ideologies do to human minds is not simply to empty them and replenish them with mistakes.

The metaphors we invoke to explain this process matter because they reflect where we place the blame. If we conceive of the mind as a container to be emptied or to be filled, someone else is doing the emptying and the filling up. The believer's mind is not complicit or responsible. Harmful ideas have been tipped into it by an external force—*The internet! Those shady friends! That charismatic preacher!* The victim is not at fault. They are under occupation.

Such images make it easy to brush away a relative's racism or a son's sexism or a friend's misinformed views. Our dearest are the unwitting fools, unknowing vessels for "bad" ideas. "We all have relatives who have blind spots," noted the poet Terrance Hayes, "but we can't erase our relatives." In moments of exasperation or paranoia we may want to shake our fanatical friends, flip them upside down, plot interventions, send them articles and gifts, hope to lure them to spit out the poison and drink our own righteous elixir instead.

In these metaphors, bad ideologies do not leak into or taint the human core. There is a possibility of innocence and redemption. Something sour and unpalatable may have taken up residence in our beloved, but the rest of them—*we are sure*—is intact, unblemished.

Metaphors are powerful because they have the flavor of explanations. But an analogy taken too literally becomes a source of error and confusion. In fact, a metaphor mistaken for the truth is worse than an error: it reinforces certain superstitions and faulty excuses. A wrong metaphor discounts behavior that ought to be scrutinized and changed. As the linguists George Lakoff and Mark Johnson wrote in their influential study of metaphors, *Metaphors We Live By*, "metaphors may create realities for us, especially social realities. A metaphor may thus be a guide for future action. Such actions will, of course, fit the metaphor. This will, in turn, reinforce the power of the metaphor to make experience coherent. In this sense metaphors can be self-fulfilling prophecies."

One of the most popular images paints the ideologue as a mindless mind. This metaphor was brought to prominence by the political thinker Hannah Arendt, who studied the trial of the high-ranking Nazi commander Adolf Eichmann for his role in engineering the Holocaust. In describing Eichmann, Arendt argued it was "sheer thoughtlessness" that drove him: he simply "never realized what he was doing." Eichmann was characterized by a "total absence of thinking," an "extraordinary shallowness," she wrote. For Arendt, what typifies the person capable of following and leading a monstrous ideology is an "inability to think."

As a result of this nonthinking, a leader wishing to instill a new doctrine would "need no force and no persuasion—no proof that the new values are better than the old ones—to enforce it," according to Arendt. When the totalitarianism of Nazi fascism was substituted by the totalitarianism of Soviet communism in East Germany following World War II, Arendt believed the overnight switch was easy. A shift from Hitler's virulent racial hierarchies to Stalin's coercive universalism was hardly registered. It is easy to conquer the comatose, the barely conscious. The citizens hardly detect the difference, the stark reversal in values, the loss of their freedoms to another regime.

Hannah Arendt envisioned the metaphor of the "absent-minded" evildoer to be literally true. It allowed her to explain parsimoniously the complex phenomena of ideological conversions and reversals. With no thinking taking place, it becomes simple to excuse thorny historical episodes or a beloved's ideological betrayal. A missing mind is easier to deal with than a mindful mind, aware of its behavior, conscious (even proud) of its crimes, and capable of behaving otherwise.

Arendt's interpretation bewitched some readers—the Nazis' actions could be explained by a kind of sleepwalking—and bewildered others. For many Holocaust survivors, it belittled the methodical cruelty of the fascist regime: the violent brutality, the systematic persecution, the careful planning of genocide. And there was another deceptively attractive implication in Arendt's metaphor of the mindless mind: it made brainwashing a simple process. A situation.

Situational explanations assume that, under the right—or wrong—conditions, we are all potential tyrants. And if we are all the same, all at fault by virtue of an inherent wickedness of the human heart, then none of us is responsible. Conformity and obedience become impossible to resist. Failures of moral conduct hinge on strong situations, insane situations, acute situations, in which all of us would become divorced from ourselves and capable of committing acts of cruelty.

In the 1950s and 1960s, social psychologists became infatuated with these situational explanations. Famous psychologists placed innocent university students in situations of torture or imprisonment; ridiculous (and recklessly unethical) experiments in which participants were drummed up to obey, conform, and inflict pain on strangers. At Yale, Stanley Milgram encouraged participants to administer electric shocks to their innocent peers—and a majority complied. At Swarthmore, Solomon Asch measured whether participants would conform to the majority opinion even when it conflicted with their own experience—across multiple testing rounds, a majority conformed at least once. At Stanford,

Philip Zimbardo split students into two groups and rendered half of them prisoners and the other half prison guards—under the instruction to dominate and exert control over the prisoners, many students turned sadistic and many suffered.

All these crazed experiments sought to prove that it is the situation, not the person, that mattered. Millions of psychology textbooks have disseminated these results and crowned the conclusions as evidence that we are all prone to conformity. In pressurized situations, we would all obey. If it is the situation that determines us, there is no need to look inside the black box, this mindless machine that we call the human being. A person can't help but surrender to the situation.

Metaphors shape the experiments we run, the theories we curate, and ultimately the moral weight with which we judge ourselves and others. If we prioritize the situation as the primary force of ideological action, we neglect individual differences and variation in behavior. After all, hidden within these famous studies purporting to demonstrate universal obedience and unwavering conformity are substantial individual differences. In Milgram's electrocution experiments, 66 percent of people inflicted the maximum level of pain on strangers when ordered—but 34 percent did not. In Asch's conformity experiments, 75 percent of participants at some point conformed to the majority opinion when they believed it wrong—but 25 percent did not. In Zimbardo's prison experiment, some participants displayed signs of rebellion and others felt so distressed by the experiment's conditions that they feigned mania in order to leave the study.

What characterizes the sizable minority of individuals exhibiting resistance? What happens inside the brains and bodies of those who reluctantly obey? If we end our explanations of evil or ideological behavior by pointing to The Situation, we fail to ask these questions. We invoke explanations of "blind obedience," "uncritical conformity," and "mindless passivity," and we neglect the evidence suggesting that there are much deeper and more nuanced processes at play.

A mindless mind cannot be held to account. A mindless mind is not responsible for its lies or fabrications, for its misbehavior or confabulations, for its harassments or its crimes. Mindlessness is not a mechanism we can measure or assess—it assumes that there is no mechanism taking place. The appeal to mindlessness as the root of evil distracts us from the quest for a scientific explanation for how minds are altered by immersion in totalizing ideologies.

The view of the ideological mind as passive and complacent, at the mercy of uncontrollable forces or evolutionary norms, is both pessimistic and wrong. In fact, the "mindless mind" is an oxymoron: What mind can be empty of itself? The brain is not an apathetic organ, inert and amnesiac. It is never an absence or a lack. Even the sleeping brain is working hard and imagining alternative realities. Comparing ideological thinking to an absence of thinking not only removes responsibility from the thinker—this account also grossly misrepresents the inner workings of the human brain.

Many of these metaphors of the mind as a container—whether full of misinformation or empty of substance—reflect an outdated assumption that thoughts are nonphysical, immaterial. This assumption is called dualism, and it implies that brains and bodies are made of different and separable substances. The body as physical and tangible and the mind as nonphysical and spiritual. Subscribers to dualism believe that the stuff in our heads—our thoughts, our mind, our personality—is fundamentally different from the stuff that constitutes our bodies—organic matter governed by chemistry and physics.

The division of the self into mind and body was famously articulated by the seventeenth-century French philosopher René Descartes. He had hoped to balance the medical understanding of the human body as a law-abiding machine with Catholicism's insistence on an immortal, incorporeal soul. In contrast to the modern framing of the "mind-body problem," Descartes initially denied there was a problem at all. "The soul and body are two substances whose nature is different," he wrote in his 1641 *Meditations on First Philosophy*, but these two substances "are still able to act on each other." Yet Descartes struggled to tackle the question of *how* a soul that exists nowhere in space could affect a body, a thing defined by the very fact that it takes up space. The philosopher and princess Elisabeth of Bohemia challenged Descartes in their correspondence, highlighting that if the soul is intangible, it cannot—by definition—have contact with physical bodies.

Descartes was in a bind. Either render the soul a matter of physics and let psychology flow from physiology (and with it, renounce the specialness of the Christian soul), or find a way to uphold the mind-matter distinction and save Christianity from contradiction with reason.

Descartes probed the brain for a location where mind and matter could alchemically interact to allow the soul's will to enter the body. A unitary location—a single heavenly gate—seemed necessary to unify the corporeal body with the ethereal nature of consciousness. Descartes knew that most brain structures exist doubled: one in the left hemisphere and one in the right. But there is a rare exception: a little organ called the pineal gland, which is a single structure, with no twin or shadow. For Descartes, this would be the seat of the soul. Despite his dualism, he was ironically tired of slippery doubles.

Dualism is the only way to believe in the supernatural and in the notion of the immaterial soul. Because only if our mental life is in some way nonphysical can we reach an afterlife or communicate with invisible entities. To believe in spiritual immortality is to believe that we are not entirely physical, biological beings. This is not only an abstract metaphysical position; it also has political implications. Ideologies that seek to separate us from our bodies often invoke dualistic assumptions. Any ideology that demands intense self-sacrifice advertises a utopia that a part of us can reach—our soul, our descendants—but that our full and fleshly bodies cannot. Religious orthodoxies and totalitarian regimes insist that we are under constant moral surveillance by all-seeing forces. Such moral scrutiny and omniscient supervision is also implied in many forms of secular righteousness about the correct way to exist in the world. To be a "good" person means to assiduously comply with the maxims of altruism, with no lapse or failure. Although there may be no one watching us as we incur personal costs in the name of a public good or a collective myth, we feel the presence of an imperceptible judge, always hovering in the background, watching over our shoulders, making pronouncements about our moral worth—the secular version of a metaphysical soul.

In these settings there is a disunity between mind and body that ideologies use to keep adherents servile and that encourages followers to commit violence against themselves in the name of a transcendent ideal.

To understand the ideological brain, we must shed these dualistic assumptions. The brain is composed of the same material as the rest of the body—water, protein, fat, salt, blood vessels, tubes, and fibers. The brain, like the rest of us, is an organ of escalating levels of cellular organization, with links to the heart and gut and our bent little toes.

Every thought we experience is physical, every emotion is physiological, every dream and conviction is a biological signature produced by the brain and body.

Yet even thinkers and scientists who deny mind-body dualism and recognize that mentality is a product of complex biology are often committed to a residual dualism. This is the insistence on a difference between the mind and the brain. In this mind-brain duality, the mind is above biology: the brain is the physical organ and the mind is the psychological experience. This partition can be useful. But the mind-brain distinction can also sustain the illusion that our psychology is imbued with a nonphysical spirituality, something eternal and otherworldly. An essence independent of the body.

Here, the boundary between mind and brain is purposefully dissolved. I use mind and brain interchangeably because there is little scientific evidence that a human mind exists that is not a brain. Change the brain—injure it, nurture it, mess with it—and our psychic lives change too.

And beyond these questions of definition (and imagery), I talk of the ideological *brain* because I want to make it clear that the mental is biological, and the biological is shaped by the political. I'm not calling to abolish the terminology of minds, brains, and bodies, but I do wish to trouble it. What happens when we remember—in our language, concepts, and metaphors—that ideologies are not only abstract and collective but also somatic and individual? What happens when we visualize ideologies as *of* the body and *in* the body?

This new science of ideology seeks to chart how ideological convictions emerge from biology. In fact, when the word "ideology" was coined, its purpose was to be a science. This history lingers in the name: ideology as a science, the *logos*, of ideas.

At a time when ideology is seen as antagonistic to science, this history now seems distant and improbable. How could ideology have ever been a matter of science rather than politics? Yet within the biography of ideology we can dig up traces of how language, history, and science intersect and meander around each other in unexpected ways—so that what begins as a science is later shelved as a relic of political history and centuries afterward can be resurrected as an entirely new scientific endeavor.

When the word "ideology" was first uttered on this earth, it was destined to be a natural science, like biology or chemistry, that investigated the nature of ideas—where ideas come from and to what ends they lead us.

How did the original meaning of ideology as a science get so radically, rudely, horribly twisted?

Part II

OF MINDS AND MYTHS

4

THE BIRTH OF IDEOLOGY

In 1794, a man sat in a dark French prison cell and wrote. Scribbling manically, his once coiffed and curled hair now in scrawny disarray, he knew his days were numbered. "Madame Guillotine" was waiting. He could almost hear the menacing grunts of his masked executioner, who would soon be nudging him closer to the guillotine's mocking blade. But instead of melancholy or a debilitating fear, he felt an electric, buzzing anticipation. *Why would a man on death row feel such pulsating energy?* The excitement was dedicated to the ideas he had discovered, ideas he felt compelled to jot down out of fear that they wouldn't complete the journey from his visionary head onto the page before his impending end.

Ideology was born this way—in the shadows of an imminent death and the longing for freedom.

The man who gave birth to ideology was unaccustomed to the brutalities of prison life. Prior to his incarceration, Count Antoine Louis Claude Destutt de Tracy was interwoven into the fabric of French nobility. Free to pursue his whims and passions, Tracy was a classic Enlightenment *philosophe*—a public intellectual committed to philosophical and political progress through reason, science, and education.

His days involved frequenting plush salons where writers and scientists debated the dismantling of France's aging feudal systems while sipping champagne and sucking on glossy strawberries. It was highly in vogue for the French *philosophes* to mock inequalities while reclining in the high-ceilinged rooms of inherited châteaux. *In vain, they tried to overlook the glaring hypocrisy.* This sprawling collection of intellectuals diligently obeyed the maxim that in order to think, one had to be deeply resourced, moneyed, and spoiled. To work hard—as they say today—one must play hard too.

For a nobleman such as Count Destutt de Tracy, prison was a serious fall from grace. During Robespierre's Reign of Terror, Tracy was captured and charged with the crimes of "aristocracy" and "*incivisme*," insufficient civic patriotism. As punishment, the count was remanded in one of Paris's most bloodstained and diseased prisons, condemned to the protracted torture of awaiting a show trial whose outcome was known in advance—a sharp and violent death.

Yet rather than mull over his misery or dwell in despair, it was here, on death row, that Tracy's imagination attained true originality. Locked up, he was also locked away from the clichés of academic institutions and their narrowing, copycat effects. Perhaps his creativity had been previously occluded by the opulence of having too much—in his former life, a starving artist he was not. Alternatively, maybe nightmares of his dislocated head rolling away from his body compelled him to find distraction. Either way, reading philosophy became his solace. Tracy devoured entire oeuvres, fixating on single authors and reading all they had published. Knowingly or unknowingly, Tracy was mimicking older generations of philosophers and poets who had been imprisoned and soon discovered they could hug philosophy for comfort, as a means of transcending their physical suffering. The sixth-century Roman statesman Boethius began the trend, composing *On the Consolation of Philosophy* while in captivity in 523 AD. Boethius imagined that Lady Philosophia—the female incarnation of philosophical wisdom—visited him in his cell. Reminding Boethius of the eternal freedom of thought, Lady Philosophia "wipe[d] his eyes that are clouded with a mist of mortal things" and dried his "eyes all swimming with tears." Bathed in relief, Boethius found that "the gloom of night was scattered, / Sight returned unto mine eyes." Lady Philosophia consoled him as a mother would and left him a clear-sighted man, able to face his bondage with bravery and peace, through the power of philosophical meditation.

A thousand years after Boethius's epiphany, the imprisoned poet Richard Lovelace lyricized in 1642 that "Stone walls do not a prison make, / Nor iron bars a cage." For Tracy too, steel bars were no deterrence. On the contrary, shackles sharpened his sense of what was most important, most urgent. The threat of death was the perfect catalyst for reimagining the power of inner thought to resist external authority and imposed superstition. As with all prison literature, the conspicuous membrane between the inside and the outside offered a vacant gap to fill with art and resistance. The cavity was an opportunity. It is therefore perhaps no coincidence that the embryo of ideology gestated in Tracy's thoughts from within the confines of prison.

When ideology was first conceived, she didn't resemble ideology as we know her today. Ideology was not a bloc of political opinions clustered together. Neither was it a doctrine that rewarded believers and punished detractors. Ideology was not a mere synonym for political beliefs.

Instead, when Tracy coined the term "*idéologie*," he wanted to understand how people came to have ideas in the first place. His aim was to introduce a new discipline, a new science to rival chemistry and physics and botany. It would become a branch of zoology, the study of animals and animate life, and borrow methods from physiology. With the formality of Greek scholarship, Tracy christened it the *logos*—the study, the logic, the rationale—of *ideas*.

Freshly baptized, the newborn *idéologie* radiated a revolutionizing potential.

Tracy envisioned *idéologie* as a legitimate science that would use objective methods to ascertain how humans generate beliefs. There would be two methods: sensation and deduction. To uncover where ideas come from, Tracy believed the science of ideas would need to pay attention to how the human mind observed and absorbed its environment—sensation!—and how it rationally formed thoughts and detected truths—deduction! "The only good intellectual mechanisms," Tracy proclaimed, are "observation and experience to gather materials, deduction to elaborate them." No other method for discovering knowledge would be good enough—"leave all the other [methods] to the pedants and charlatans," sniggered the *philosophe*.

Today we might find parallels between Tracy's musings and modern experimental psychology, the study of cognition, of how we perceive, judge, and act. But at the time of Tracy's theorizing, the term "psychology," with its "psyche"—the contemptible "soul"—sounded too spiritual for the secular thinker. His aversion to religion meant that discussions of the soul were out of the question. The spirit evoked the divine, and there was nothing mystical or otherworldly about truth. Tracy was adamant that his science of ideas would be a solid science, not philosophical abstraction. The titles "psychology" or "metaphysics" would be misleading misnomers. There was no room for theology in science. The human intellect had a biological basis.

By assuming this perspective, the inspired count was rejecting Descartes's mind-body dualism and religious inclinations. Instead, he was following in the footsteps of the seventeenth-century empiricist philosopher Francis Bacon, who ardently believed that it was necessary to "re-make entirely the human mind, begin all the sciences again, and submit to a new examination the whole of our acquired knowledge." *Throw out all that is implausible, faulty, erroneous, unjustified, taken for granted! Only unfiltered truth will remain.* In accordance with Enlightenment ideals, Tracy contended that there should be a reevaluation of how knowledge was acquired and that any acquisitions that could not be justified by experience or reason be discarded.

So when ideology was born into this world, into the prison's dense air, it was designed to be a scientific project oriented toward truth and against superstition. Like a parent gazing lovingly at their baby, Tracy felt certain that *idéologie* would be beautiful and "sublime." She would elucidate the power of human intellect and observation, reveal the elements of sensations, humanity's cartwheeling capacity for truth. (Tracy had a taste for the melodramatic.) And although *idéologie* was designed to be a secular, scientific endeavor, it was not going to be devoid of morality. Ideology would shine a light on which ideas were grounded in firsthand experience or robust reason and which were the flimsy and untrustworthy inheritance of religious orthodoxy. Through the science of ideas, Tracy believed we could create methods that would clearly distinguish truths from falsehoods. Above all, ideology would be *useful*: it would "show the human intellect the road to be taken to increase its knowledge, and to teach it a sure method of reaching the truth."

On the twenty-second of July 1794, the moment arrived to distill Tracy's insights into a final manifesto—one that would outlive him. The day had been one of the summer's hottest, temperatures rising and agitation swelling. Perhaps the accumulated heat trapped in the prison cell melted some shyness or reservation in Tracy's heart. He was finally prepared to conclude his truths. The next day, as he sat in his dingy cell in Paris's prison des Carmes, the soft urgency that had enveloped Tracy's incarceration grew louder. Time was up. Earlier that morning, nearly fifty of Tracy's fellow inmates, some similarly accused of the crime of aristocracy, had been shackled and led to meet their end. As the echoes of the guillotine's pitiless chops reverberated in the distance, Tracy knew that the Revolutionary Tribunal would announce his own verdict within a week.

"Summary of Truths," he titled his diary entry on the twenty-third of July. Truth, for Tracy, demanded the highest respect, the utmost objectivity. His conclusions, he decided, ought to be formalized through the vocabulary of mathematics. But since Tracy was not a trained mathematician, pseudo-algebraic equations would have to suffice. He began:

"The product of the faculty of thinking or perceiving = knowledge = truth."

Tracy's first axiom implies that rationality and observation are the routes to knowledge and hence to the truth that resides behind reality. If one reasons clearly and perceives the world sensitively, misleading veneers will splinter and shatter. So far, so Enlightenment, so good.

"Three other terms must be added to this equation," he wrote next, almost greedily, on a fresh line:

"= virtue = happiness = sentiment of loving."

In the second axiom, Tracy continued the first, proposing that truth leads to goodness and joy. Even the slippery emotion of love. A little ironic for an incarcerated and solitary *philosophe* wasting away in the bleak midsummer of a Reign of Terror. But even in his limited state, trapped in a dark cell, Tracy championed the virtues of closely attending to the senses. In his notes, Tracy had even tried to riff off René Descartes's famous "I think, therefore I am" and ambitiously hoped to recast it as "I sense, therefore I exist." For Tracy, our existence is defined by our sensorial experiences, not merely the presence of inner thoughts. Like mindfulness, Enlightenment-style, Tracy believed that sensitive perception would lead to happiness. Matter and mind were not separable for Tracy—the Cartesian dualisms of body and brain, biology and thought, were misguided categories. Our minds are continuous with our world. If you experience sensory impressions—the heat of the sun on your cheek, the texture of rough paper under your fingertips—you will prove your existence through the act of sensation.

(Sadly for Tracy, his reformulation of Descartes's ode to pure interiority didn't catch on. It didn't have the same melodic ring as the original.)

As Tracy's epiphanies coalesced on that hot July day, channeled through this final "Summary of Truths," the mathematics of reality felt vivid, almost fluorescent. Like the aura caressing old Boethius's face when Lady Philosophia visited his cell like an apparition, Tracy reported a "shaft of light" beaming with insight.

To complete his formula, Tracy added one last set of variables:

"= liberty = equality = philanthropy."

From his first equation to the last, Tracy deduced that truth is anchored in our direct observations and capacities for reason. When these are exercised, the trail leads to goodness and happiness, and ultimately to freedom, justice, and altruism.

It's a long equation.

An equation that all too enthusiastically relies on the equal sign to leap between big declarations, somersaulting from the human faculty of perception all the way to a magnanimous love for humanity.

What was the meaning behind this algebraic soliloquy? Were these the musings of a madman? A prisoner experiencing grandiose and megalomaniac delusions? Someone hoping for a sense of significance or salvation—of affirmation that his lone thoughts were linked to higher existential and political ideals? Possibly.

But Tracy should not be simply dismissed as deranged. His philosophical calculus suggested that by unlocking the *origins* of ideas, one could illuminate the *consequences* of ideas. Tracy understood that our sensorial perceptions may be linked to our visions of equality and emancipation.

As the French Revolution mutated into the Great Terror of the 1790s, the Enlightenment as a historical period was beginning to fade. Tracy wanted to extend the Enlightenment into the nineteenth century, carrying its spirit kicking and screaming on his aching back, campaigning for reason and self-governance over monarchy, slavery, censorship, and feudal servitude. The European Enlightenment philosophers who preceded Tracy had taught him that progress involved shedding the skins of childhood fantasies—or, as the philosopher Immanuel Kant put it, emerging out of our self-imposed immaturity. To think for oneself, to independently search for truth, leads to both intellectual and political freedom.

Would the truth, or at least its wholehearted pursuit, set Tracy free?

The hours leading to his unjust trial were ticking by impatiently. With little time left before the executioner's grip would tighten around his resisting arms, the count might have reasonably supposed that *idéologie* would die prematurely, still in the cradle, never having been caressed by the white Parisian light.

Tracy had no faith in divine intervention, but he must have read and reread his "Summary of Truths" like a prayer. Did Pascal's famous wager cross his mind? Maybe beginning to believe in God, in the supernatural, would be a wise and harmless bet. Believing in God was like a gamble that would either evaporate upon his death if God did not exist or guarantee a life in heaven rather than hell if God was indeed real. The philosopher Voltaire, Tracy's childhood idol, on whose lap he once sat in awe and adoration, had declared the wager's reasoning to be "indecent and childish." Yet not all idols could afford to be decent and mature. Some faced an upcoming death by guillotine.

In a breathtaking historical coincidence, Tracy was granted a last-minute date with fortune. That same week, a twist of fate would transform Europe and save *idéologie* from the chopping block. Four days after

Tracy penned his concluding diary entry—his truths dutifully summarized and dated—the French parliament revolted against the Terror's ruler Maximilien Robespierre, who was captured on July 27 and jumped the queue to the guillotine. The Reign of Terror ended the following day. Tracy's life was spared. The Revolutionary Tribunal intended to judge Tracy's innocence or culpability never went ahead. By October, Tracy was released and returned to family and friends—and within a couple of years was sitting back with other surviving *philosophes* in lavish salons, writing his magnum opus *Elements of Ideology*, and introducing the world to the vision of his newfound science. One of Tracy's most enthusiastic readers, an American named Thomas Jefferson, quickly volunteered to translate the text into English and popularize it widely.

Perhaps truth does liberate after all.

As Tracy carried *idéologie* out for public display, arms full, a proud parent ready to boast and peacock, he quickly amassed both allies and enemies. In his role as a politician and policymaker with a penchant for secular education, Tracy wanted to position his vision in every house of learning. If every school and university instructed its pupils in *idéologie*, in the methods of close sensation and analytical deduction, the next generation would achieve true enlightenment. Instead of relying on convention or descent, young students would center their beliefs on reason and observation—on a robust understanding of epistemology. Future citizens would become unshackled from traditional mores and airless doctrines. Education would be a tool for political liberation.

So although *idéologie* began as a plan for a new empirical science, once it was exposed to the open air it oxidized and quickly changed hue, becoming a political mission. Many of Tracy's contemporaries supported the utopian dream of a society geared toward cultivating individuals' mental faculties and capacities for critical and lucid thought. These intellectuals and policymakers banded together and called themselves the *idéologistes*.

"Ideology would change the face of the earth," remarked one of Tracy's students after a conversation with the energized ideologists. "That is exactly why those who wish the world to always remain stupid detest ideology."

But who would want to keep people foolish and stupid? What kind of person craves an ignorant society?

5

THE AGE OF ILLUSIONS

As the mist began to lift in post-Terror France, a heavy-booted rising star on the political scene by the name of Napoleon Bonaparte heard about the ideologists' new movement and thought it sounded good. Or, at the very least, useful. The ideologists could serve as a gateway to power—their social connections and influence could become essential for traversing France's new political landscape. Napoleon began associating with the *idéologistes*, attending their events, circling their parties, and affiliating with their organizations. Raising glasses and toasting to Tracy and his friends, Napoleon flattered and befriended, and soon received an honorary membership to the ideologists' institute.

But, like most seductions, the flirtation ended almost as soon as it began. Differences in values swiftly appeared. The ideologists were committed to democratic, secular, and libertarian principles, keenly celebrating the power of the individual mind. They cheered at the idea that the French nation would represent the voice of the people instead of religious authorities or royal elites.

The ideologists' aims, however, were woefully incompatible with Napoleon's vision. The Machiavellian manipulator had imperialist ambitions, planning to reinstate clerical institutions and become the architect of a tightly controlled dictatorship. Minds were meant to obey, not to express doubt. Minds should not frolic freely. They must be contained.

For Napoleon, the most convenient strategy for dealing with Tracy's club would have been to dismiss or ignore the movement. Wave them away disdainfully as he consolidated his clout. The problem was that Tracy and his crew were not simply armchair metaphysicians, tucked away in libraries and exclusive philosophical circles. Annoyingly for Napoleon, they were no longer physically imprisoned. Instead, the ideologists were active, powerful, and deeply networked legislators. Napoleon had tried to send Tracy away on an expedition to Egypt, but Tracy wittingly turned down the invitation. If you can't beat them, thought Napoleon, mock and curse them.

"Yes, they are obsessed with meddling in my government, those windbags!" Napoleon barked in 1802, like a petulant child. "My aversion to this race of *idéologues* amounts to disgust."

In jeering disrespect, Napoleon twisted the term *idéologistes* into *idéologues*, and the word "ideology" became an insult, a slur, flung scornfully at political opponents. Napoleon labeled Tracy and his cohort as abstract idealists with no practical skill or sense. "They are dreamers and dangerous dreamers," snapped Napoleon at anyone who cared to listen, trying to belittle the ideologists as pretentious, hubristic, and out of touch.

By 1804, a decade after Tracy's imprisonment, the militaristic Napoleon was more than just an ambitious politician—he was France's authoritarian and dynastic emperor. "It is the doctrine of the ideologues," Napoleon trumpeted, "to which one must attribute all the misfortunes which have befallen our beautiful France." The fight garnered such publicity that Madame de Staël, a witty French political theorist exiled for her critiques of the Reign of Terror and

Bonaparte's tyrannical personality, pronounced that Napoleon suffered from "ideophobia." *Fear of ideas, fear of ideology, fear of logic.*

Napoleon rabidly foamed and declared: "Gentlemen, philosophers torment themselves to create systems; they will search in vain for a better one than Christianity, which in reconciling man with himself assures both public order and the peace of states. Your ideologues destroy all illusions, and the age of illusions is for people as for individuals the age of happiness."

The age of illusions is the age of happiness. Illusions are virtuous. Illusions are *necessary*. Those seeking to break illusions are the enemies of the people. The real world is too harsh a place; people ought to be hypnotized for their own good. It is only through the enchantments of illusions that people will unify into a devoted community, a well-ordered gelled majority.

Tracy's war with Napoleon left Ideology bitter, defeated, and sidelined. Rejected and rendered the scapegoat of French society's ills, Ideology traveled abroad. Fleeing persecution and disgrace, it sought solace in America. But the Americans had also heard of Ideology's tattered reputation. Even Tracy's friend and translator Thomas Jefferson, the third president of the United States, couldn't undo the damage. In letters and meetings, some of America's Founding Fathers derided ideology as a "science of lunacy," a "theory of delirium."

Ideology had committed the crime of centering reason and observation at the expense of collective myth and magical thinking. The Americans believed it lauded the individual—the possibility of the rational, independent, enlightened citizen—to the verge of psychosis. The newly formed United States relied too heavily on aspirations of communal identities to host Ideology as a concept in its freshly declared territories.

Disappointed, Ideology crossed the Atlantic back again to the European continent. It landed in Germany, in the lap of two young thinkers who were in deep discussions over the nature of illusions. Yet, unlike Napoleon, they did not *marvel* at illusions. Oh no, they were in the business of smashing them.

By the time Karl Marx and Friedrich Engels came to write *The German Ideology* in 1846, "ideology" had metamorphosed again. It was no longer simply a pejorative accusation directed at political rivals. By the middle of the nineteenth century, in the hands of Marx and Engels, ideology constituted the beliefs internalized by human minds through coercion and exploitation. Ideology was far from science. Ideology was the set of illusions themselves.

Writing of religion as ideology, Marx argued that "to abolish religion as the *illusory* happiness of the people is to demand their *real* happiness." Ideological illusions were false solutions to miserable and unequal living conditions. "Religion is the sigh of the oppressed creature, the heart of a heartless world, and the soul of soulless conditions," Marx wrote in 1843. "It is the opium of the people."

Mockingly, Marx called Tracy a "fish-blooded bourgeois doctrinaire." Although Tracy invented and nursed ideology, Marx claimed that Tracy had entirely missed the point. Instead, Marx and Engels turned to one of Tracy's contemporaries for inspiration, the *philosophe* Claude-Adrien Helvétius. "Our ideas are the necessary consequences of the societies in which we live," Helvétius had suggested in 1758. The notion made Marx and Engels giddy with a subversive kind of frenzy. The ideas in our heads are *impure*. Forget Tracy's optimism about the potential of reason and observation to yield reliable truths! Individual consciousness is weak and faulty: it submits to the structures and tricks of society. Ideas are inherited and bendable, easy to seduce and maneuver. There is no innate human logic or human perception that can be dissociated from culture, capital, or power.

Helvétius had written of the ways in which the ruling strata of society—the royal aristocracy, the military, and the clergy—were keeping dominated groups in the dark about the human condition. Through propaganda, disinformation, and manipulation, rulers were dictating the beliefs of the masses. The young Marx and Engels loved this theory. Parroting and maybe even plagiarizing Helvétius, Marx and Engels famously wrote that the "ideas of the ruling class are in every epoch the ruling ideas."

The duo suggested that ideological myths serve the interests of the few, not the many. People are largely ignorant of the social and economic relations shaping their thoughts and desires, suggested Marx and Engels. "In all ideology," they wrote, "men and their circumstances appear upside down as in a camera obscura." Consciousness is "inverted" by ideology. Everything is turned on its head. Up is down and down is up. Human beings tolerate and even defend systems that ultimately impoverish them. Consciousness is social, interlocked with class and capitalism—it is something to be contorted and controlled. In conditions of inequality, mental resources are sucked from subordinate groups, and illusions are transplanted in their place.

Marx and Engels believed ideology could be equated with *false consciousness*: "phantoms inhabiting the human brain."

For them, ideology situates us in an *Alice in Wonderland* kind of world where distortions are everywhere. Appearances and reality are mixed up and confounded. Social constructions are mistaken for permanent structures of reality. And importantly, we forget that these distortions exist. We see impenetrable fortresses where crumbly sandcastles stand. The fragility of our dogmas is cloaked by an impression of strength and longevity, so that we forget that dogmas can be broken.

Marx and Engels sought to show that within an ideology, inequalities are made to seem natural, immovable, inevitable, and good. Imbalanced power relations are naturalized, as though social life was always this way and always will be this way. The scripts we follow feel to us as immune to change or challenge. We might as well play along. *Who are we to defy nature?*

Under the spell of false consciousness, people take on duties and self-definitions that are given to them, imposed upon them from outside, which they have no hand in designing or actively consenting to. Citizens become instruments of authority, of history, alienated from themselves.

Only through a drastic shift in society's structure, only by allowing individuals to determine the conditions of their lives, will people be truly liberated from the illusions that bind and blind them. Piercing through facades is the only way to come into possession of true consciousness. Only by shattering the veneer of ideology can we wake up.

Whether we understand Marx's claims or not, agree or disagree, feel armed with insights and ready for the revolutionary battle or disenchanted by the conspiratorial and populist overtones in Marx's philosophy, there is something intoxicating about the desire to cut the strings that sustain the puppetry of modern inequality. Let it all fall apart and a new age—an age *against* illusions—be born.

Is that applause? An evil laugh in the background? The crackling fires of the destruction necessary for renewal? Or a muted silence? *What is your response?*

The calamitous irony in Marx's hatred of ideology is that his own ideas collapsed into an ideology in itself. Marxism became a narrative of supposedly indisputable logic, with clear prescriptions, an outgroup to hate (the "ruling class"), and an ingroup to adore and glamorize (the "ruled"). You either bought into the logic or you were a traitor, as bad as the oppressors.

In response to Marx, a vicious tug-of-war over the concept of ideology ensued. It has lasted two hundred years and is still ongoing. Gallons of ink have been poured over the question of whether ideologies are a help or a hindrance. Some claim that ideologies are useful because they provide organizing frameworks that render life intelligible and social exchange possible. Others suggest that ideologies stimulate self-deception and entrench inequalities between groups. Patriarchal ideologies lead to gendered myths about the self—including one's desires, capabilities, and needs—that people accept and live by, allowing injustice to persist in plain sight. Monarchist ideologies foster close feelings of kinship with symbolic aristocrats that blind people to the blatant exploitation that enables these monarchies to survive. From the perspective of the critic, the belief that these inequalities are justifiable and good is a symptom of ideological illusions.

The debate over whether ideologies enable or disable human life continues to rage. Ask your neighbors and some will worship ideologies, calling them necessary and beautiful principles that mark a moral man, distinguishing him from a bad one. Others will spit in disgust, cursing ideology as the barrier holding people back from their authentic possibilities.

Historically, defining ideology has been both a descriptive project and a political, normative one. Whether we see ideology as a force for good or evil (or both or neither) is entwined with the definition of ideology we espouse. Whether we believe ideology operates at the individual level or the social level (or both or neither) affects where we think it exists and whether it is possible—or desirable—to detach from ideological frameworks.

Whereas Tracy glorified individual cognition, Napoleon mocked it. Marx thought individual consciousness did not exist. For Marx, consciousness was always collective: it was the product of external, material conditions. "Life is not determined by consciousness," he argued, "but consciousness by life."

Ideology is the air we breathe, Marx assumed. We cannot, we do not, live without it. We are born into the atmosphere of an ideological structure *and then* we think and exist. Ideology creeps into and conditions our entire mental life. Ideology, for Marx and Engels, *precedes* human consciousness. Our inheritance of morality, religion, and social scripts comes first. Our chosen lives come second. Ideology defines us before we have the chance to define ourselves.

Yet Marx's formulation leaves almost more questions than answers. If ideology is so pervasive, can we ever evade its grip? Is it possible to dispel the poisonous fumes while preserving the oxygen that sustains us and keeps us grounded within our societies? And if it is—somehow—possible to escape these pernicious systems, what is the way out?

In the midst of all these unsettling and complicated questions lies the suspicion that even if we manage to see things as they "truly are" we might not be able to bear the world we find. Once we peel ideology off the surface of things, would there even be a solid, objective reality out there? Would we know how to live or whom to trust? Is it possible to divorce from our ideologies without falling back into an ideology again?

In interrogating the nature of ideology, it is easy to lose our way—to enter a recursive skepticism that leaves us in a state of vertigo, a tired restlessness, a desire to let all these questions remain unanswered and declare "enough!"

Out of the mob of questions that have been solicited of ideology throughout the nineteenth and twentieth centuries, multiple definitions and theories grew. The study of ideology relocated from a premature empirical science to the domain of the humanities, where the phenomenon of ideology spread into every corner of what it means to be human. Scores of thinkers dedicated their lives to these tangles. It obsessed the nineteenth-century titans G. W. F. Hegel, Friedrich Nietzsche, and Sigmund Freud, as well as the twentieth-century icons Jean-Paul Sartre, Albert Camus, W. E. B. Du Bois, and Michel Foucault. The question of ideology galvanized the civil rights and feminist thinkers whose work motivates the "raising of consciousness" in the twenty-first century, including Simone de Beauvoir, Frantz Fanon, Martin Luther King Jr., and Audre Lorde. The biggest names in philosophy have grappled with ideology, trying to hold it by its collar and force out a confession. *Who. Are. You?*

The challenge has been that the definition of ideology inflated, expanding beyond recognition and beyond measure. Everything was ideology. Ideology spilled into the concepts of domination, power, and culture. We could point to any action, any thought, any social relationship and be pointing, angrily, at ideology. It was a fat target on which any arrow would land.

The concept of consciousness experienced a similar definitional expansion to the one that befell the notion of ideology. The meaning of consciousness was stretched and kneaded like dough until it was only a semblance of the original phenomenon. Some definitions of consciousness looked outward at the moment-to-moment sensations and impressions received from outside, while other definitions gazed inward, construing consciousness as an awareness of a self that is thinking and existing consistently across time and space. In this reflective—almost recursive—formulation, to be conscious is not merely to sense or to think but to be aware that one is sensing objects or thinking thoughts. Consciousness entails a kind of self-witnessing.

In almost every language, "consciousness" has come to possess multiple meanings: consciousness as awareness, wakefulness, sentience, reflection, and even political conviction. Linguistically, this mixed baggage is no accident. The first philosophers to delineate consciousness as a phenomenon of interest were seventeenth-century French philosophers like René Descartes and his followers, who borrowed the Latin term *conscientia*, which originally meant "conscience" in the moral sense and only later came to be understood as "consciousness" in the psychological sense too.

But even after moral awareness and mental awareness became differentiated and separable concepts, the idea of psychological consciousness continued to be a source of contention. Consciousness cleaved into subcategories and branched off into contradictory definitions. Many thinkers doubted that consciousness even existed in the first place. At the end of the nineteenth century, the pioneering psychologist William James famously saw our subjective life as consisting of a "stream of thought"—a rich succession of feelings and sensations—which has been overwhelmingly (mis)remembered as a "stream of consciousness." James was doubtful that a unitary consciousness could be found or that it was a useful concept. "I believe that 'consciousness' . . . is on the point of disappearing altogether," he wrote. "Those who still cling to it are clinging to a mere echo, the faint rumor left behind by the disappearing 'soul' upon the air of philosophy. . . . It seems to me that the hour is ripe for it to be openly and universally discarded." To James, "consciousness" was perfumed with a latent spirituality, a thing that we claim exists by virtue of our introspective feelings but for which there is little concrete evidence. Such philosophical debates on the nature of psychological consciousness occurred in parallel to political debates on how social consciousness becomes warped by capitalist and religious myths.

Soon, everything was either within our consciousness or *un*consciousness or *sub*consciousness. Consciousness could be split or bifurcated. As early as 1897, the scholar and civil rights activist W. E. B. Du Bois—who was William James's philosophy student at Harvard—wrote of "double-consciousness" as typifying the African American experience: "It is a peculiar sensation, this double-consciousness, this sense of always looking at one's self through the eyes of others, of measuring one's soul by the tape of a world that looks on in amused contempt and pity. One feels his two-ness."

Being aware or awake or woke became a riddle of levels and tiers. Herein we find the precursor for the twenty-first-century terminology of being "woke." Psychoanalysis and political activism began to consider how consciousness could be raised or elevated or even weaponized in order to move people away from self-denial and toward enlightenment about themselves and their surroundings. Consciousness lost its boundaries, flowing out of the human body and onto the streets. Everything was everywhere, all the time. When everything is everywhere, maybe it is nowhere to be found.

With all these myriad definitions, emphases, and nebulous limits, it may be that the surest way to study the links between ideology and consciousness is by thinking about them as constructs to be measured. A definition that is rooted in measurement is a more promising starting point for conversation than a definition rooted primarily in abstract theories, which may at times be untranslatable and incommensurable, allowing clever people to speak past each other.

It is here that the experimental science of the conscious brain becomes an unlikely hero. Two hundred years after Tracy's hopeful musings, modern science can rescue ideology and consciousness from their boundless expansion into everything and nothing. When ideology and consciousness are denoted and tested as cognitive phenomena, it becomes possible to investigate the parallels between people's sensory consciousness and their political consciousness—how their phenomenological experience of the world is colored by their ideological convictions about the world.

Tracy's vision was of ideology as a science; a rigorous method for discovering when ideas are faulty and unreliable. Yet this vision was abandoned and the psychological project was forsaken.

The umbilical cord tying Tracy and *idéologie* snapped apart. The loving labor invested into its conception in the early years was now lost and misappropriated. Count Antoine Destutt de Tracy became a forgotten mother. It's no surprise—we rarely thank our mothers enough.

Unrecognizable, like a released prisoner traumatized by war, the definition of "ideology" began to drift—pushed around haphazardly by the waves of history. Battered and bruised, ideology shifted its entire essence. We could not remember where ideology began and where culture or criticism or politics ended.

Yet to repel the magnetism of myth, I believe our best bet is to return to empirical science. Perhaps a second look at the human mind—how it thinks, how it perceives, and how it acts—can reveal what an ideological mind does. What an ideological mind desires or deflects. What an ideological mind destroys. How an ideological mind can be destroyed from within.

To understand the dangerous liaison between minds and myths, we need to understand what propels each of them. What are their personalities and preoccupations? Why are they so easily enticed by each other? Should this toxic affair be stopped? How do ideologies snake into our fast-learning, dogma-loving brains—and what are the consequences? Can ideological indoctrination ever truly be reversed or erased?

Let's begin with the magnificent, the fallible, the looped and coiled human brain.

6

BEING A BRAIN

One should be skeptical of any story that crudely divides an idea into two: light and dark, heaven and hell, good and evil, love and hate. But, embarrassingly, the human brain seems to be largely characterized by *two* fundamental principles. Of course, it has many more functions and properties. Certainly more than two. Otherwise we would be a wildly boring species. Yet much of human behavior can be accounted for by these two characteristics, and so they are useful and illuminating for mapping out what a brain does and how it can be led astray.

The first principle of the human brain is that it is a predictive organ. It learns associations from the environment and tries to envision the next occurrence. Like an ambitious weather forecaster, the brain picks up patterns from its past and hopes to anticipate the future. It is constantly trying to understand the world—to have an accurate representation of what is out there and what is about to happen. By perceiving billions of subtle connections and coincidences, such as what happens when we drop a suitcase or step on a banana, the brain builds an internal model of the world. This model of the world will be governed by our intuitive understanding of physics. Gravity will lead the suitcase to fall to the ground, the dubious solidity of the banana implies that if stepped upon it gets grossly squashed—and as a result, someone will have to clean up the mess. Chains of cause and effect become aggregated into an internal representation of the world around us, and give us an orientation for what (or who) to chase and what (or who) to avoid.

Nearly three centuries ago, the Scottish Enlightenment philosopher David Hume observed how sensed associations of cause and effect come to guide human behavior. He wrote in 1748 that "it is certain that the most ignorant" among us—which Hume prejudicially thought included "infants," "stupid peasants," and "brute beasts"—"improve by experience, and learn the qualities of natural objects, by observing the effects which result from them. When a child has felt the sensation of pain from touching the flame of a candle, he will be careful not to put his hand near any candle; but will expect a similar effect from a cause which is similar in its sensible qualities and appearance."

It is only a foolish man or a cheeky adolescent who will return to touch the candle's flame after it has burned them earlier.

Babies develop a sense of the world through millions of experiences and form implicit *expectations*, predictions about the effects that follow an event or an action. Developmental psychologists can test infants' evolving abilities through experiments that violate babies' expectations—where instead of falling down, an object magically flies upward or sideways when it is released from someone's grasp. If a baby has learned that objects tend to fall down, they will be surprised if they see an object fall straight upward. Their intuitive predictions will be breached. As a result, infants will tend to watch unusual or physics-defying events for a longer time than mundane movements; their brains will be hard at work trying to reconcile their expectations with their new observations. This may be partly why young children are so mesmerized by simple magic tricks. These little scientists are naturally forming hypotheses, feeling reassured when the hypotheses are confirmed by the incoming data, and becoming intrigued or alarmed when their prophecies fail.

By the sheer power of experience and repeated exposure, the brain learns what is physically possible and physically impossible. It witnesses the world, acts upon it, and constructs theories about its machinations and hidden relations. Similarly, the brain processes an enormous number of patterns regarding social interactions, learning the rules of communication. This allows it to predict what is socially possible and socially impossible. The brain forecasts what might happen in any given scenario. Physically, we may predict that stepping on a banana can lead to an uncomfortable physical *squish*. Socially, we anticipate that rapidly evacuating the scene without clearing up the freshly created mess will be frowned upon and may be punished. We can err in our predictions—*maybe instead of receiving disapproval, someone will sympathize with our plight*—but the brain is perpetually theorizing and prophesizing nonetheless.

Somewhere between learning about physics and understanding social relations, we also learn about power dynamics and the limits of our agency. For toddlers, when an object falls to the ground, someone typically picks it up and returns it to them. Gravity seems to surrender to the toddler's desires. So children can develop a warped expectation that takes decades to unlearn—that someone else will undo their mistakes or will bend over to return their prized possessions on demand. Many continue feeling entertained and entitled to these games of power and powerlessness into adulthood. Those around them, backs aching, tend to suffer. The expectations we generate—about others or about ourselves—do not occur in a vacuum: our predictions are contingent on the reactions our actions evoke from the world.

So a fundamental guiding principle of the human brain is that it is a generative and creative predictor. It picks up on patterns easily and quickly, and intuitively extrapolates these impressions to the next event. The human mind sees narratives everywhere—sometimes even in places where they do not exist—filling in the gaps to achieve coherence and meaning.

Importantly, the brain is not satisfied with a fictional model of the world; a model based on fantasies and miscalculations. In building a mental representation of reality, the brain seeks to maximize accuracy and to minimize its errors.

The brain is hungry for the truth and (generally) nothing but the truth.

Why the stubborn insistence on accuracy? What is so bad about mistaken predictions?

Like Indiana Jones, the mind is both a professor and an explorer. Theoretical and practical. The brain is not merely an academic generating hypotheses from the comfort of an armchair or the seclusion of the ivory tower; the brain is in the field. It is constantly prepared for action. The brain has evolved to move and respond to its surroundings, to be alert to risks and hurdles. Every observation is a source for opportunities and affordances.

Since the brain is fundamentally designed for action, it is geared toward speed as well as accuracy. In order to respond quickly and efficiently, the brain learns from past experiences to extract rules for behavior. It searches for simple instructions and no-fuss, straightforward formulas. Easy heuristics that it can follow instead of complex correlations and hidden connections that the mind must laboriously discover and compute. Adhering to a parsimonious rule that captures known regularities—expect gray clouds to signal oncoming rain, anticipate a hug after a reconciliation—is more efficient than figuring out contingencies every time from scratch. It is tiring to discern nuance and endless possibilities at every crossroad.

Figuring out the rules that underpin our physical and social lives helps the brain fine-tune behavior and control its surroundings. After all, knowledge is mastery and mastery is power. Accurately modeling the world and its particularities is an excellent strategy for successfully living in the world, helping us to complete the fundamental tasks of self-regulation and survival in multidimensional environments.

Some people affectionately call the brain a predictive *machine*, to express the impressive automaticity, precision, and intensity of its forecasting skills. But, luckily, there are many differences between a motor and a mind. Whereas machines are inanimate instruments, cogs and pegs, ones and zeroes, human minds are alive and organic. Brains comprise billions of cells that sit in complex and hierarchical arrangements in the spongy tissue inside our heads, and are woven together elegantly and richly with the rest of our bodies. As "biological engines" composed of cellular matter, brains are subject to more randomness, noise, and stochasticity than electric or analog engines. The interactions between cells are more uncertain and indeterminate than mathematical computations that always yield the same answer. A calculator will never stray from the equation 5 + 4 = 9. But a human adult counting on their fingers will sometimes accidentally respond with "8" or "11." Biology is more fallible than pure mathematics. Our brains are full of flukes and quirks.

Yet with all their idiosyncrasies—and maybe *because* of the all-too-human oddities resulting from such patterned mistakes—human brains are also incredibly powerful. When computer scientists train modern machines to compete with humans on seemingly simple tasks and games, the power required to generate such calculations is so immense that running these models is eye-wateringly expensive and environmentally harmful. Huge quantities of electricity must be channeled into these computations, illustrating that what a single brain does effortlessly is profoundly difficult to re-create and imitate. Simulating the predictive and interpretative powers of the brain—how it can forge coherence out of chaos and link together events into layered stories—is now a multibillion-dollar industry. The race is on to design complex algorithms that capture the human capacity to learn, generalize, and create causal narratives with fidelity. At the moment, the gap between human and artificial intelligence is sufficiently wide that we can still rely on crudely cropped images of bridges and traffic lights to signal that, yes, we are human.

The brain is actively and dynamically predicting the world and its inhabitants—including itself. By mapping the probabilities of events in the world and updating expectations with every confirmation or refutation, we construct a replica of reality inside our heads. This representation of how the world is and how we imagine it will be in the future allows us to navigate our lives peacefully and, hopefully, profitably.

But the story does not end here. The brain has another characteristic that—when knotted with its predictive capacities—brings out the best and the worst in us. It is the double-edged sword that is at the root of the tenderest kiss and the most heartless war.

This second principle is that the brain is fundamentally communicative. Participating in social life is essential to survival and reproduction, and so the brain wants attention. It yearns for reciprocity, for the undulating back-and-forth of acknowledgment and connection.

Like many animals, human brains are oriented around relationships. Dogs sniff each other in initial suspicion and later play, just as we humans assemble first impressions and laugh about them afterward. As whales sing to each other below stormy seas, we lull each other with lullabies or whistle to catch attention from afar. A pair of shiny frogs idly sharing each other's company on a wet lily pad are reminiscent of two elders relaxing together on a park bench. A plethora of animals use visual and acoustic displays to attract mates and advertise their sexual availability or kinship relations. They groom and soothe, tumble, play, and run in circles. Some even dance, performing and posturing like matadors or macho men at the gym. Like us, animals are constantly communicating, signaling their presence, availability, and needs.

When in distress, animals and humans alike will try to comfort, to touch, to embrace. We coo and aah at images or videos of toddlers consoling a crying parent, or horses bending over to nuzzle a fallen rider in pain. The adult-like caring capabilities of young children or anthropomorphic tendencies of animals often enchant and impress us.

Some of these overlaps are superficial and accidental—it is unclear that frogs enjoy sharing lily pads or care about the company. (Yet several studies show that male frogs' vocalizations to attract female mates depend on the presence of rivals posing competition, so maybe frogs do care about the nature of the company.) Other parallels between the communication styles of human and nonhuman animals are based in deeper structural similarities. These resemblances—deep and shallow—attest to the impulse to come into communion with others—to disclose and discourse, to be in relation. "All real living is meeting," poeticized the philosopher Martin Buber in his 1937 book *I and Thou*. "Where there is no sharing there is no reality."

The word "communicate" originates from the Latin verb "to share"—and when we speak, gesture, or listen, we engage in the act of sharing attention. We greet by shaking hands, bumping fists or elbows, bowing heads in synchrony, feeling our faces come together nose-to-nose or cheek-to-cheek. To trusted friends, we open up, we make ourselves physically and emotionally vulnerable when we begin an exchange.

To mark how and with whom we should enter conversation, humans develop rituals. Social rituals establish trust by demanding that we come into communion. Oaths of allegiance, a scout's salute, proudly sung anthems, the daily adornment of sacred symbols, totems, and piercings. To constitute effective signals of communicative trust, rituals must be transmissible and exclusive. We develop visual, musical, linguistic, and tactile signatures to signal *This is me! This is us! And no one else!* Like code words, rituals signify a shared and unique identity. *If I move like you, I must be like you, and so maybe I should like you and invite you in.* Rituals facilitate synchrony: they force us to move in unison, to coordinate our thoughts and feelings, to repeat the idiomatic mantras that become the basis for our shared reality. When we chant together, sing together, rhythmically march or dance or sway together, pray or om or exhale together, the lines between us become fuzzy. After participating in collective gatherings or social movements, many individuals report: "I felt bigger than myself." Their eyes sparkle from the memory of how skins seemed to melt and bodily borders became more porous. Together with others, the individual feels bigger, mightier, better.

The sociologist Émile Durkheim endowed this emotional energy with the name "collective effervescence" in 1912 to capture the vivacious aura of communicating as a group. No one is immune to the enlarging and euphoric effect of raucously applauding for a musician's encore—*One more song! One more song!* We become united in a shared mission, a joint purpose. Our movements are mirrored back to us, swelling and swallowing us up. Gaining confidence, we march more boldly, sing more loudly, dance more enthusiastically, as though infected by a new kind of joy or passion or peace. Our desires merge and multiply; the resulting wave is stronger than the sum of its parts.

Solidarity is contagious. Our social brain longs to feel part of a story, part of a group. When minds communicate, they transcend the confines of skulls and skins. No one is alone. Everybody can be understood. Everything is shared.

The human brain's communicative abilities culminate in the motivation to feel attended to. To feel at home. To feel belonging. To feel that our lives have significance, that we matter.

For the brain, we *must* matter, our subjectivity must count; otherwise the model of the world we have created—in which we act, predict, explain, and anticipate, correct our expectations and communicate frustrations, tell stories and build identities, tolerate complex and at times awkward social lives—all this predictive and communicative labor will be in vain. We must matter for all the psychological burdens to have value and meaning. A brain's life must be worth living.

The brain seeks comprehension and recognition—compensation for the energy it requires to decode and participate in social life. The brain examines and wishes to be (gently, kindly) examined in return. After all, as Socrates supposedly said, the unexamined life is not worth living. And the blunter French-Algerian philosopher Albert Camus famously claimed that "there is but one truly serious philosophical problem, and that is suicide."

So, in order not to die, not to kill or be killed or kill ourselves, the brain's attention to the world must be accompanied by the feeling that the world is paying attention to it too. We must feel that we understand the workings of the world—we need a reliable representation of our reality—and we must feel understood by others—we need communication.

Existence teaches us that our actions are followed by rippling reactions. We learn to expect to be heard when we speak, to reciprocate when we receive, to be answered when we ask.

Ideologies are the brain's delicious answer to the problem of prediction and communication. Ideologies provide easy solutions to our queries, scripts that we can follow, groups to which we can belong. Guiding our thoughts and actions, ideologies are the shortcuts to our desire to understand the world and be understood back.

But many perceived shortcuts accidentally lead us down a longer road. Most solutions entail new problems. An easy fix rarely exists. The ancient Greeks recognized this conundrum and invented the word *pharmakon*—the linguistic ancestor of the pharmacy—to denote both a cure and a poison. After all, medicine both helps and harms. A drug can sometimes be a remedy and at other times a toxin. Often it is both simultaneously, producing painful side effects and nauseating reactions.

Perhaps ideologies are sophisticated *pharmakons*: resolving the brain's problems of prediction and communication, but also creating fresh complications, potentially more severe than the original condition.

Emulating the rigor of the doctor—as well as the prospective patient's intermingled hope and anxiety—we can ask how ideologies are different from other stories we tell ourselves about ourselves and our surroundings. To diagnose the effects of ideologies, we ought to confront ideological thinking in all its bittersweet glory and explore how the brain's antidote to its dilemmas can become its tragic undoing.

7

THINKING, IDEOLOGICALLY

Nowhere else is life so neatly ordered as from the inside of an ideological doctrine. While it may be a historical accident that "logical" sits within the word "ideological," it is a telling coincidence. Ideological thinking is true to its name: it is *super*-logical, *hyper*-logical, and that is why it is so tantalizing and so dangerous. It is not irrationality that propels a person toward ideological thinking—it is the desire for a perfect and foolproof logic. "Ideological thinking orders facts into an absolutely logical procedure which starts from an axiomatically accepted premise, deducing everything else from it," Hannah Arendt rightly observed. "That is, it proceeds with a consistency that exists nowhere else in the realm of reality."

An all-powerful premise claims to explain everything. It is a theory of everything. Everything can be predicted and explained. The past, the present, the future. Our existing conditions and impending prospects. The origins of our disquiets and frustrations. How we ought to act and with whom we must never interact. Why humanity has struggled and how the struggle can stop. Who deserves blame and who deserves praise. All the "hows" and "whys" and "whos" are answered definitively.

In an ideology, life is determined. Determined in the sense of *discovered*: the truths of the past and future, the path to moral goodness, have been logically deduced or mystically found. And life is determined in the sense of *decided*: life can adopt no other course and is destined to be so and so and not otherwise. Each individual fits into the cosmos in a particular way and this destiny cannot be denied or averted. At least, not without punishment.

Regressive ideologies that look backward nostalgically—seeking to preserve or reinstate old hierarchies of power, rooted in characteristics of gender or geography or race or class or caste—exercise control over their followers with the threat of violence or material deprivation (another kind of violence). Progressive ideologies often aggress against detractors more indirectly, more obliquely, than regressive ideologies. But in progressive ideologies too, the person who rejects the inevitability of the progressive utopia and fails to act consistently on its morality at all times can be deemed as deserving of attack and disgrace. Acting badly or inconsistently—affiliating with the wrong group, consuming the wrong products—is an irreversible stain. The ideologue, whether regressive or progressive, will find ways of appointing people to binaries of good and evil, with nothing in between or beyond.

To think ideologically is to see morality as immovable. Any changeability is seen as suspicious. There is an assumption that people do not truly change and only fools deny what is predetermined.

All ideologies are stories of inevitability, but is the reverse also true? Are all stories of inevitability also ideologies? Perhaps every kind of causal or moral inevitability—even if it has no official label or name—becomes a kind of determinism. In science, when an expert delivers a prediction with no range of uncertainty, it begins to resemble a prophecy rather than an estimated probability. Within the sacred entity of the family, when a couple *must* stay together despite all obstacles and misalignments or when a child *must* respect an abusive parent despite the parent's repeated cruelties, an inevitability is invoked again. In social life, when a person feels imprisoned by a duty, by an expectation that they *must* and *should* be a certain way, despite the discomfort that such conformity may precipitate, there is a rigidity at play. There is a contraction of the future and an abolition of alternative possibilities. A denial that things could change or be different.

It is difficult to find a kind of determinism that isn't also a latent dogma: a dogma-in-waiting.

In *The Origins of Totalitarianism*, Arendt delves into the all-too-consistent and all-too-competitive thinking of totalitarian regimes. Nazi totalitarianism emerged from a fixation on life as a fight between races, and Stalinist totalitarianism was built on life as a fight between economic classes. The explanatory premise of most ideologies claims that life is governed by a fight between groups, such as a battle between nations, between economic classes, between genders, between races, between nature and humanity, or between divine gods and earthly apostates. By following the ideological premise to its logical conclusion, entire systems of beliefs, moral codes, and protocols can be erected.

Without doubt or hesitation, ideological premises describe the state of the world and prescribe actions that would usher in a better version of that world. An ideal world. A utopia to counteract the current condition, because the present—for an ideology—is depicted either as a horrid dystopia or as an optimal state that must be preserved and protected from looming threats. Ideological reasoning oscillates between triumph and victimhood, depending on which of them stimulates devotion and action more effectively.

If the ideological premise dictates that life is a fight between groups, then everything is perceived as an existential struggle for scarce resources, a battle for domination and self-determination. It is a zero-sum game. All actions that lead to victory are legitimate. This is one of the most fascinating features of a feverishly defended ideology: the way that everything becomes permissible in order to attain the ideal. If a person has internalized a nationalistic doctrine in which there are essential differences between people residing on either side of a border, then a potential national threat appears to excuse the killing of innocent civilians. If averting an ecological collapse becomes the sole existential concern, then there is nothing too drastic, no sacrifice too excessive, to fight on behalf of the environment—innocent lives can be endangered if it directs people's attention toward the havoc of climate catastrophe. The ideological premise explains why an action is necessary and excuses its unpalatable dimensions. If one follows the urgent logic, all actions that lead to victory are valid. An ideology begins to be practiced extremely when the premises justify even the most radical of means.

This does not mean that contradictions do not bedevil ideological thinkers—they certainly do! Any simplifying account of reality is rife with inconsistencies. The ideological thinker downplays the importance of counterevidence, trying to compartmentalize it away. But counterevidence does not simply bounce off the skull without entering the brain. Indeed, faced with counterevidence, the ideologue is rarely dispassionate or apathetic—they are typically annoyed. They are aggravated precisely because they must deal with an irksome dissonance, because cognitive effort is required to suppress the predictive brain's truth-seeking and truth-tracking tendencies. Although expectations structure human perception and interpretation, the brain can still register and remember counterevidence. Implicitly recorded inconsistencies can wait silently, patiently, in the background. It is often in a moment of crisis—existential, political, familial, financial—that contradictions can percolate to the surface and belief change or moderation can occur. Until then, for the committed ideologue, there is a logical sweep that glides over everything.

Totalitarian leaders have been acutely aware of the "irresistible force of logic"—in Stalin's words—that governs ideological rhetoric and that "like a mighty tentacle seizes you on all sides . . . and from whose grip you are powerless to tear yourself away." Slick and slimy, the tentacle hugs us tightly and hooks us in. We hug the tentacle back, clinging on firmly, nestling in. The embrace comforts us. *We don't want to resist.* We press ourselves against it, feeling ourselves cradled and wrapped by the force of reason. *We sink into it.*

There is a sense in which ideological narratives do the two things that brains seek to do: predict and communicate. To our certainty-seeking brains, a systematic theory of everything sounds splendid. To our community-loving brains, a shared theory of the world seems fantastic. When we look at the internal structure of ideologies, we see that there are affinities between the properties of human cognition and the properties of ideologies. Ideologies seem to possess two essential qualities: a *rigid doctrine* and a *rigid identity.*

First, all ideologies embrace some form of rigid doctrine that assumes the existence of one true explanation of, and corresponding solution to, existing societal and existential conditions. Ideologies offer absolutist descriptions of the world and accompanying prescriptions for how we ought to think, act, and interact with others. A doctrine organizes life into fixed moral categories: the categories of good versus evil, truth versus falsehood, right versus wrong. The doctrine rationalizes prescriptions and prohibitions, punishments and rewards. By virtue of its perfect logical systematicity, the doctrine is considered beyond doubt. "Ideologies are sealed universes," wrote the cultural theorist Terry Eagleton, "which curve back on themselves rather like the cosmos, and admit of no outside or alternative." Ideological principles and policies repel evidence or interrogation. The credibility of counter-evidence is challenged or its conclusions are deemed so preposterous and offensive that the implications are considered impossible. I call this the *doctrinal* dimension of an ideology: the evidence-resistant absolutist descriptions and prescriptions that guide the believer's thoughts, actions, utopian hopes, and imagined catastrophes.

An individual's level of dogmatism reflects how rigidly they hold on to beliefs and how hostile they feel toward alternative opinions. The dogmatic individual is punitive toward dissent or contradiction and is unlikely to shift their beliefs when the evidence commands it.

Dogmatism has a doppelgänger, an angelic twin, the inverse of closed-mindedness. A trait called intellectual humility. If a person is intellectually humble, they are open to revising their beliefs in light of credible evidence or strong counterarguments. They are open to plurality in debates, accepting the presence of different viewpoints and perspectives. They do not feel personally under attack when someone disagrees with them. Their intellectual ego is not fragile and permanently on alert for threatening counternarratives.

In contrast, if a person is low on intellectual humility and leans toward dogmatism, they are more hostile to nuance. They prefer clear-cut solutions and absolutes. They believe in generalizations about how the world is and ought to be. They avoid navigating the labyrinths of murky evidence.

An individual's dogmatism may originate from a kind of intellectual *servitude*—a tendency to submit to the majority opinion and defer to others' authority rather than engage in independent reflection. A kind of chosen intellectual diffidence or uncritical persuadability. Let others decide and lead the way.

Alternatively, a person's dogmatism may stem from an intellectual *overconfidence*—an arrogant conviction that their way of thinking is superior to all other ways of thinking. They have special access to the truth. Perhaps by virtue of a rare and prophetic personal intelligence, or through deep philosophical contemplation and scientific investigation, or due to the extraordinary luck of belonging to the right family, the right religion or political sect, they are uniquely positioned to recognize and relay the truth. They see themselves as "genealogically lucky," in the framing of philosopher Amia Srinivasan; the stars aligned just right so that their birth positioned them in the arms of the correct worldview. Unlike others, the unlucky majority born into the wrong family or the opposing tribe, they are blessed.

Ideologies enforce a sharp distinction between those who are in possession of the ideology's truth and those who are not. This is the difference between followers and nonfollowers, believers and nonbelievers, the ingroup and the outgroup. And so in addition to a rule-bound dogma, there is always a social element to ideologies: a division between those who belong to the ideological ingroup—"us"—and those who do not—"them."

As a result, adopting a rigid doctrine is frequently intertwined with embracing a rigid identity.

This second feature is commonly achieved by inventing distinctive identity markers, such as flags, symbols, songs, anthems, costumes, and rituals, which signal membership and devotion. The shared and visible nature of these identity markers fosters passionate feelings of immersion and connectedness with the ideological group. Indeed, people will kill and die over a torn flag or a defaced ideological symbol.

The language of kinship and familial relatedness is frequently conjured by ideological movements. Metaphors portray comrades as "sisters- and brothers-in-arms," religious leaders as "mothers and fathers," the nation as the "motherland" or "fatherland," and revolutionaries as the "sons and daughters" of ideological causes. Emotionally, ideologies seek to mimic the sense of belonging, familiarity, and intimacy of home, engendering all the loyalties and self-sacrifices we would make for our family.

Through these exclusive identity categories, nonadherents are rejected or shunned, and become the objects of hostility and prejudice. To ensure obedience and commitment to the group, ideologies demand participation in repetitive as well as one-off rituals. These rituals can be costly, humiliating, or physically painful. Traumatic rituals that permanently modify the body have always been used by ideological groups, large and small, to foster and prove membership. This is manifest in initiation rites and hazing rituals, spiritual piercings and tattoos, religious circumcisions and mutilations, arduous pilgrimages or military service, voluntary incarceration and self-negation in the form of fasting, dieting, or vocal silencing. Almost all ideological groups establish and test identity by rituals that involve the shedding of blood or the self-infliction of pain.

Ritualistic behaviors are hard to fake or pursue halfheartedly; the suffering is too great to bear if it is not accompanied by conviction. This makes them optimal identity markers to designate who we should consider a stranger or a sister. I think of this as the *relational* dimension of the ideology: the strong favoritism toward fellow adherents and antagonism to nonadherents that can lead to distrust, prejudice, discrimination, and ultimately violence.

To be ideological is not as simple as extolling an *-ism* or identifying with a particular group. To be ideological is to adopt a rigid doctrine and a rigid identity. To *think ideologically* is therefore to think in a way that rigidly adheres to a doctrine and resists updating beliefs in light of new evidence, as well as to think in a way that rigidly attends to the identities of ingroup and outgroup members. Both the doctrinal and relational dimensions are necessary for ideological thinking, and neither alone is sufficient.

Can a person have a rigid doctrine without a rigid identity? Or a rigid identity devoid of a rigid doctrine?

In principle, yes. People can take on rigid identities without espousing rigid doctrinal beliefs. Ardent supporters of sports teams or musical superfans are intuitive examples. It is common to feel a strong sense of affiliation with fellow devotees of the celebrated group or artist and to despise and demonize anyone who feels otherwise. Many exclusive social identities are built on such mutual interests and passions—whether in sports, music, literature, fashion, or the high school clique. In these cases, people are socially categorized and divided and at times prejudged or discriminated against, but there is little to no clear doctrine or a fully fledged belief system underpinning the rigid identity.

The reverse exists too: rigid doctrines without accompanying rigid identities. It is possible to believe stubbornly in ideas without layering them with social identities. Such dogmas are systematic and resistant to evidence but do not explicitly make social claims about the virtues of believers or nonbelievers. Economic and technological doctrines can take this form, such as when they make definitive pronouncements about how to achieve human prosperity. For instance, economic neoliberalism can be practiced as an explicit or implicit dogma that holds certain assumptions about the causal links between governmental intervention and public outcomes or the moral worth of people and money. It dispenses prescriptions about the operation of free-market capitalism and government regulation, but it does not seek to build family-like relationships between believers. Adherents to economic neoliberalism will rarely identify themselves as such, unless they are seeking election to public office. Economic doctrines do not always directly deal with personal selfhoods. These are doctrines that do not necessarily invoke explicit social identities.

There are also more amorphous examples that hang between these poles, such as privately practiced rigid behaviors. Extreme diets or exercise regimens in which there is a dramatic reduction or increase in food intake can become—if sustained over a long period—compulsions that are difficult to stop. These can flower into private, nonpolitical kinds of dogmas, with personal rules that must be followed and feelings of guilt when there are lapses or perceived failures. The individual becomes their own censor and regulator.

All these instances of doctrine without identity or identity without doctrine fail to capture the psychological phenomenon of ideological thinking in its entirety. In order for an individual to exhibit ideological thinking in the full sense, I believe they need to internalize both an absolutist doctrine and an inflexible social identity.

Importantly, although it is possible to have a rigid doctrine without a rigid identity, and vice versa, there are many instances when one kind of rigidity leads to the other. An economic neoliberal ideologue who begins to threaten nonbelievers, exhibiting discrimination and hostility to detractors, *is* ideological in the full psychological sense. A person who becomes obsessed with a diet and begins to evangelize it to others, to judge and shun those who question it, to inflict harm on themselves and others in the pursuit of the diet's goals *is* developing a rigid identity twinned to their rigid diet dogma. A sports fan who wears the team's colors and acts aggressively toward the fans of other teams and at the same time resists evidence that questions the team's technical supremacy, even when the team fails again and again in front of their eyes—yes, maybe the fan is developing a kind of evidence-resistant doctrine that structures their decisions and skews their observations. Many rigidities are on the cusp of morphing into an ideology.

There are gradients to ideological thinking. An ideologically extreme person is one who adopts the rigid doctrine passionately, leading them to possess an absolutist, evidence-resistant description of the world, and to strongly adhere to inflexible prescriptions for how they and others ought to live and act. The ideologically extreme person will also embrace the rigid identity passionately, leading them to exhibit intense identification with fellow adherents and display active malice toward nonadherents.

Defining an ideologically moderate individual—the weak partisan, the occasional believer—is a more nuanced endeavor. Ideological moderation can come in different configurations and in varying intensities of dogmatism or devout identities. The portrait of the ideological moderate depends on whether we believe they exist on the opposite pole to the ideological extremist or whether we feel that ideological moderation is a stepping stone toward a nonideological position that antagonizes ideological thinking at every opportunity. Thinking about the nonideological person who resists strict doctrines and identities is one pathway to understanding the other side of the spectrum. Doctrinally, a nonideological person tends to adopt a description of the world that is flexible and responsive to evidence and does not rely upon or impose on others rigidly prescriptive rules for living. Relationally, they display weak identification with those who believe in similar worldviews, and do not express hatred or prejudice toward others who do not. The nonideological person strives toward intellectual humility—continuously being open to updating their beliefs in light of credible evidence and balancing a healthy dose of skepticism against mythmaking practices with a humanist sympathy toward those who feel compelled to engage with collective ideologies.

We live with many ideologies within and around us. Some become salient, beloved, and intense. Others wait in the background for moments in which to become expressed—competitive events, national referenda, encounters with people or ideas considered foreign. When a person exhibits thought patterns and responses of a dogmatic and intolerant nature, we need to pay attention to *how* they are thinking and not only *what* they are claiming to fight for.

The question of whether a believer's ideologies concern race, gender, class, climate change, religion, nationhood, or politics is irrelevant to whether the person can be designated as ideologically extreme or moderate. By focusing on the structure of ideological thinking rather than its substance, this approach can help us identify which individuals are ideologically extreme and which individuals are less ideological in their outlook.

This framing is resilient in the face of slippery relativism. Relativist positions assume that we cannot judge or compare different cultural or ideological practices because each is unique. Yet unless we can describe the phenomenon of ideological thinking in a way that is agnostic to the content of the beliefs, we will struggle to critique and identify oppression or call out extremism in public life. Without a sense of what ideological extremism looks like regardless of its mission, we will struggle to refute demagogues who label basic human rights as tyrannical or use "culture" as a shield for injustice. *After all*, they can say, *it's all relative*.

Without clear definitions, tolerance starts to fold in on itself—tolerating intolerance until the intolerance dominates again. Diagnosing extremism objectively is essential for understanding the past and engaging in the ideology wars of the future. By differentiating ideological and nonideological thinking and acknowledging the intermediate grades and shades, we can develop a better understanding of how ideologies can be opposed within a humanist framework. As Steven Pinker advises in *Enlightenment Now*, we should not fall prey to the notion that "we might as well give up on reasoned persuasion and fight demagoguery with demagoguery."

By concentrating less on whether an ideology is true or not, whether it is illusory or illuminating, we cultivate a more palpable sense of the psychological texture of ideological thinking, of what it means to be immersed in an ideology regardless of its sociological history or affinity to the left or to the right. This understanding of ideological extremism also allows us to discern the emergence and development of ideological extremity within a person—how and why indoctrination can begin early and spread beyond political beliefs and moral imperatives and into the brain's *non*political life.

In the 1954 book *The Nature of Prejudice*, the social psychologist Gordon Allport wrote that "a person's prejudice is unlikely to be merely a specific attitude toward a specific group; it is more likely to be a reflection of his [or her] whole habit of thinking about the world." It is therefore possible that our political prejudices and ideological convictions are particular instances of more general habits of thinking.

Exploiting the human passion for clarity and connection, ideologies magnetize the entire brain. Dogmas are the sexy seducers whispering in one ear: *Want to know what to think?* And into the other ear: *Want to know who to love? Who to hate? Where to go from here?*

Let's see who will follow. Not everyone is easily seduced.

Part III

ORIGINS

8

A CHICKEN-AND-EGG PROBLEM

The eleven-year-old ponders the question posed by the professor.

"What is your country's biggest problem today?"

The child shifts in the plastic seat. Their eyes rest on the tape recorder in the middle of the table.

"Taxes on everything, and the cost of living."

Taxes, mumbles the professor, jotting down the answer in rapid pencil strokes.

"All right, thank you. My next question. How would you change your country?"

Watching the spinning wheels of the tape recorder, the child's walled eyes betray nothing. The pupils then begin to enlarge and brighten—as though struck by an epiphany or a vivid memory—and move up to meet the professor's gaze.

Confidently, the child exclaims: "Clean up the streets! All that garbage lying around . . ." The child's face contorts in disgust. "See that everything is in order!"

The eleven-year-old's words produce a soft echo in the bare room.

The professor tries to sustain contact with the child's glare, now more menacing than before. Power seems to have shifted.

Without looking away, the eleven-year-old reaches up to fix stray strands, assimilating them back in line with the gelled majority.

Eventually, the interviewer cowers down back to the notepad. *Purity and discipline*, writes the professor, *fear of chaos*. And then almost as an afterthought: *Potential fascist?*

In an adjacent room, a thirteen-year-old carefully considers the same questions.

"The biggest problem is the starving people of Europe," the child explains gently, "because the people in our country won't think of them, and they should."

Oriented toward love, catalogues the professor, *compassion*.

"How would you change your country?"

"We should have a world police," the child replies, "so that there would be no more wars."

Surprised by the child's response, and its imagination of cause and effect, of police as the antidote to violence, the professor probes. "Why?"

The thirteen-year-old pauses and contemplates. "We should be peace-loving, not out for power."

Will against *power*, the professor inscribes with a gratified chuckle. *Cooperation instead of domination*.

The professor looks up at the thirteen-year-old sitting opposite before bowing back down and writing in the margin: *A classic liberal child? . . . Or a copycat of liberal parents?*

Among the liquid syllables of liberalism, is it possible to distinguish between the independent thinker and the imitating conformist?

Enough of the interview, thinks the professor. Need to delve deeper. *Next: administer cognitive tests for an accurate assessment of unconscious beliefs*.

Thought experiment or parody? Such exchanges seem implausible. Surely children do not have political beliefs, let alone coherent ideological positions. Yet these are, in fact, dramatized sketches of real interviews—interviews that took place in 1944 between a curious professor and surprisingly articulate children. Now forgotten fragments of scientific history, the interviews relay the real words and genuine documented opinions of children who participated in a pioneering study into the elemental signs of prejudice. A study that explored how early, and in what forms, prejudice can emerge.

The researcher, Else Frenkel-Brunswik—a recently arrived refugee in America escaping Nazi Austria—led one of the largest investigations of prejudice in children ever conducted. She wanted to study how and when "the ethnocentric child becomes a potential fascist." Which children were most prone to xenophobic and authoritarian thinking? To find out, she transformed hundreds of ordinary California children into experimental participants and studied them extensively. Her provocative hypothesis: the responses of children on the cusp of adolescence might foreshadow their future tendencies to resist anything foreign and believe in the absolute superiority of their own racial and cultural identity.

Over 1,500 children between the ages of ten and fifteen were tested with questionnaires about attitudes toward minority groups that were relevant to the rapidly changing Californian context—beliefs about segregation and immigration, views about Japanese and Chinese neighbors, Jewish people, and African American acquaintances or strangers. In their responses to these questionnaires, some children exhibited an unconcealed xenophobia and others resisted clichéd stereotypes. Based on their answers, hundreds of children were subsequently chosen for in-depth interviews. The candidates were selected from two groups: the *most* prejudiced, the most hateful against minority groups, and the *least* prejudiced, the most liberal and accepting of others.

By inviting children to the interview table—the metaphorical couch in the psychoanalyst's study—Frenkel-Brunswik hoped to shed light on the development of the potential fascist. She guessed that children's reports and stories would contain insights into their political vulnerabilities. By mapping out how children resolve contradictions and confusions, when they adapt and when they stiffen, one might decipher which children may be most tempted by ideological narratives and leaders. The interviews probed the children's ideological beliefs as well as their personalities, their home lives, and how they viewed themselves and others. When they arrived in the interview room, one by one, the interviewer was blind to the children's prior scores. It was only after the interview that Frenkel-Brunswik could cross-reference the children's questionnaire scores with their verbal accounts about their families and themselves and the stories they offered when probed about political affairs.

What distinguished the most xenophobic children from the least? Could the differences between them exist not only in opinions but also in the kind of minds they possessed—the particular ways their brains processed information and arrived at conclusions?

"If a potentially fascistic individual exists, what, precisely, is he [or she] like?" Frenkel-Brunswik and her colleagues at Berkeley asked. "What are the organizing forces within the person? If such a person exists, how commonly does he [or she] exist in our society? And if such a person exists, what have been the determinants and what the course of his [or her] development?"

If such a person exists—a potential fascist, a mind amenable to trusting a totalizing ideology—what are their dispositions? Could we see the seeds of an individual's ideological potential while they are still in childhood?

As the interview data accumulated, one insight became clear: it was not difficult to distinguish between the prejudiced and unprejudiced child, between the xenophobic and liberal child. In fact, it was remarkably easy to tell them apart.

How could Frenkel-Brunswik spot the differences? Was it the children's proclamations or their expressions? Their indulgence in verbal clichés or some marker beyond language—outside the realm of conscious reports, perhaps subtle bodily movements or reactions—betraying the child with potential for prejudice before they uttered a single word? What were the telltale signs?

A Catholic girls' primary school in England recently claimed that we need to "treat the brain like a muscle, not a sponge." The report was accompanied by a cartoon of a brain lifting weights. With enough discipline, it suggested, everything is attainable for everyone. No need to worry about natural differences or dispositions and preferences. Nurture a child with enough force and all of nature will yield.

Hierarchical ideologies—ideologies that stipulate untainted reverence for a prophet, an idol, a guru; ideologies that ask the devoted to kneel, literally or metaphorically—often argue for an ironic kind of equality. *No child will be left behind.* False proclamations of sameness are not accidental. Insisting on our indistinguishability from each other can be a device for domination. If we are all the same, we can all be converted, corrected, controlled. Preaching is more difficult if it must recognize nuances or the limits of its power.

In reality, our brains do not bow to authority with equal enthusiasm. Some will acquiesce quickly, bending forward so completely their foreheads kiss the ground. Others will cave in slowly, the rib cage resisting obedience's concave movements. And others, usually very few, will be unwilling to submit to ideological dogmas—they will have to be pushed or kicked down. A firm grip will be required to hold their necks still, cheek rammed against the floor. Even then, their eyes will remain open, looking straight on in defiance.

Which brains will submit and which brains will resist?

While there is a general magnetic pull between minds and myths, between the predictive tendencies of the brain and the predictive structures of ideologies, its force is stronger for some people than others. These differences in vulnerability can be traced to the personal idiosyncrasies of our brains.

After all, brains are not born the same. Differences exist along countless dimensions to create the stunning diversity we observe between people, even within the same family. Some minds are impulsive and others are patient; some are rule loving and others are rebels at heart; some are loyal and group-oriented while their cousins are stubbornly independent. Some brains solve puzzles faster than others; some brains are cautious, taking the slow-and-steady approach. Others are fast and furious, making quick and sometimes wrong decisions. A few are sensitive, tuning in to the world with a sharp ear or taste buds that detect the slightest variation. Some minds react with great fury at the sight of injustice; other brains are less attuned to such suffering.

Like bodies—because brains *are* our bodies—brains come in different shapes, sizes, and proportions. They possess specific allergies and distinctive appetites and affections. Each brain is idiosyncratic and singularly structured and so each perceives, responds to, and learns from the world in subtly different ways.

These individual differences matter because they give us clues about how strongly a person will feel the pull of ideological doctrines. We are not equally controllable.

And now the question becomes *why*. Why are some minds at risk of ideological thinking while others are resilient? Which traits and experiences predict the substance of our beliefs and the extremity of our convictions? Is it our dispositions or our situations? Our inborn quirks or our learned habits?

To disentangle causes and effects, we need to figure out what comes first and what arrives second. Is it our brains or our politics? Our personalities or our ideologies? Is it a matter of temperament—certain kinds of brains gravitating to certain ideological allegiances—or do ideologies, passionately held, come to shape our brains and personalities?

This puzzle is a chicken-and-egg problem. It is a question of causality, of origins and consequences. It is also implicitly a question of what is fixed and inevitable versus what is flexible and reversible.

The chicken-and-egg problem of political neuroscience is born out of the fact that whenever we assess the political mind, we are getting a singular snapshot in time. We capture an individual in that one second—their ideologies and their brains at that one instant, a blink in a life. But when we take this kind of psycho-political photograph of an individual, we do not know how they got there or why. The snapshot hides information about what happened minutes before, or days earlier, or in the years and generations that preceded this hour. Is it the effect of socialization or the result of a preexisting biological vulnerability?

The challenge is to unscramble whether individuals' psychological characteristics determine their political, religious, nationalistic ideologies *or* whether ideological indoctrination has a transformative impact on our brains and bodies.

Which way do the arrows point?

Perhaps the arrows point both ways—our brains sculpt our private politics and at the same time our ideologies shape the functioning of our brains.

If our vulnerability is inborn, we need to figure out the traits that make a person vulnerable. We need to understand how our individual cognitive quirks and biological responsivities can give rise to powerful *-isms* for which we are willing to hurt and be hurt, causes for which we would kill and die.

If our vulnerability is partially acquired, and the neurobiological signatures of ideologies are gained over time, then the consequences of immersion in ideologies are more profound, intense, and life-changing than we commonly realize. Hannah Arendt, in one of her most astute intuitions, wrote that "what totalitarian ideologies aim at is not the transformation of the outside world or the revolutionizing transmutation of society, but the transformation of human nature itself." Arendt was talking about how totalitarianism tries to corrode the barrier between the private and the public, the personal and the political. But I think there is an even deeper transformation that extreme ideological systems can instigate. Ideologies do not merely wash our brains, ridding us of old notions and replacing them with new ones. Ideologies transform our cognition, our reflexes, our visceral and biological nature. Disentangling the causes and effects that lead to these patterns is the puzzle we're here to crack—where our ideological possessions come from and how deeply these social inventions change us and possess us from within.

9

YOUNG AUTHORITARIANS

Interviewing children about their political beliefs seems, at first glance, ridiculous. What can children know about world affairs? What could children possibly understand about war and domination, government and inequalities, atrocities and injustices that exist far from their doorsteps?

Ask them, thought the psychologist Else Frenkel-Brunswik. As a Jewish child in Vienna in the 1920s, Else Frenkel grew up into a world that was increasingly racist and inhospitable to people like her—a world that quickly and not-so-quietly turned from a golden intellectual age into an oppressive and violent regime. Young Else was forced to learn about war and domination, the nature of government and inequalities, atrocities and injustices that were knocking impatiently at her door. If she did not understand the serious world of adults, she would be lost. If she did not decipher adults' lies and intentions, their hushed mumbles and secret affairs, she would be in danger.

A precocious student, Else completed her psychology doctorate at the age of twenty-two at the University of Vienna and began lecturing in biographical psychology. As though foreshadowing an impending history, Else asked what psychological intimations can be surmised from biography.

Even before Hitler's annexation of Austria, Else was targeted by racist students who complained about being taught by a Jewish lecturer. After experiencing a personal interrogation by the Nazi secret police—the Gestapo—in 1938, she felt the political bleeding mercilessly into the personal. Else fled Europe with fellow psychologist Egon Brunswik and sailed with him to America. Just before touching the shore, Else and Egon married. They settled in Berkeley, California. In this new country of liberty and opportunity, the promised land of freedom, she would begin to explore the psychological reverberations of *unfreedom*.

Frenkel-Brunswik took up a research position at the University of California, where she turned to a group she believed could be astute informants on what it means to be unfree: children.

What do children know of chains? Nothing, most modern parents would say—it is the children who are the masters, the lieutenants of the family unit, tyrannically demanding more and more. Toys, gadgets, clothes, candy, privacy. They need and need and need. Need becomes a synonym for want. *Buy this! Take me here and drive me there! Now leave me alone! Shut the door!*

Yet, perhaps in the most important ways, children are the constant subjects of a totalitarian-like structure. The existential condition of being a child is akin to being in a hierarchical ideological system with constant surveillance and the need to appease all-knowing leaders. Resources are concentrated in the hands of a few powerful authority figures. For a child, life is governed by opaque rules and rituals whose meaning and purpose are often unclear and unexplained. The family is a site of fundamental inequality, defined by an inescapable asymmetry in dependence. Power, in the family, is unevenly distributed. Every decision, every plea, is dictated by the fact that the child cannot survive alone. The caregiver gives, and the gift of care is not free. In return, there is payment and debt. Children, after all, are "among the most powerless and casually brutalized creatures in the world," observes the literary critic Merve Emre. While there can be opportunities for collaboration and harmony, there are also demands for devotion and ample room for oppression. Freedom, for a child, is inherently limited.

Implicit in Frenkel-Brunswik's thinking was the idea that perhaps patterns of authority and unfreedom learned within the family—an insistence on strict obedience and conformity rather than a prioritization of care and creativity—might render some children susceptible to political and religious authoritarianisms. Children who are well versed in hierarchies of power and arbitrary violence within the home might be more likely to participate in authoritarian systems later in life. They might be more receptive to clichés. In contrast, children from families that foster imagination, empathy, and acceptance of plurality might be more resilient in the face of totalitarian ideologies. They might instinctively reject submission. Because they have been marinated in freedom. Because they can imagine otherwise.

Although some children were shy and others smooth talkers, Else Frenkel-Brunswik believed that with patience and respect it was possible to elicit answers to the most private questions, even from children who were reserved or apprehensive. In some sense, children were better test subjects than adults; they were "more direct and uninhibited" in their responses. After all, the bluntest truths—undecorated by euphemisms and diplomacy—often come from children who have not yet learned the art of self-censorship.

Else knew that to interview means to receive answers that are unavoidable blends of truth and distortion. Fact and fiction, projection and aspiration are all bound up and knitted together. Maybe at least with children as her experimental participants, she had a better chance of teasing apart appearance and reality. Finding consistencies among children's fables and folk philosophies of life somehow seemed more straightforward than extracting truths from the self-righteous stories of grown-ups. Adult artifices, twisted through retrospective excuses and unrealistic hopes, are difficult to untangle.

I would resent living next to somebody that was a foreigner," the fourteen-year-old girl announces. "They should go back to where they came from."

"Where should foreigners and immigrants go?"

The professor's follow-up question is almost a breach of protocol. It comes across as a cold rebuke rather than a neutral inquiry. *Challenge without antagonizing*. But it is too tempting to see whether the adolescent's logic would crack and break under the pressure. Whether thin fractures would materialize in the teenager's rationalizations.

The interviewee slouching in front of the interviewer appears unfazed by the interrogation.

"Foreigners should return to the country they came from. Another thing too," the fourteen-year-old adds dispassionately, "some of them were born in America, a lot of them. I still think they should go back to where they came from."

The professor suppresses a frustrated exhale and asks: "How?"

"Well, the people in our country should just make a law to chase them out when they come in."

Chase them out.

"What about displaced people whose homes were destroyed in the war?"

The girl retorts matter-of-factly: "Well, it was probably their fault that the war started." She throws blame with ease, with a kind of cool, glamorous, cruel casualness.

A piece of gum emerges visibly from behind her row of incisors. Sticking her tongue into it, the girl blows a small cherry bubble. It nearly grazes the tip of her nose.

Transfixed, the professor holds a stifled breath. Something other than logic is being stretched, pulled, and stressed.

The girl's Bazooka, the pinkish inflated balloon, finally pops.

An inaudible air puff is released.

"And in any case," she adds, "they got what was coming to them when their houses were bombed."

The next interviewee sighs sadly at the question. "Where we live now they don't allow the _____ people there."

"What do you think about that?"

"It's silly. I don't think we should have those rules. The _____ people have such lousy homes. I think they ought to have just as much right to buy property here as anybody else. They need new homes in good districts."

Anti-segregation, the professor writes on the yellow notepad, *sympathy for minorities' rights.*

"What do you think of _____ people?"

"I know a _____ boy. He is real nice. Some _____ shouldn't even be called _____ because they are awful nice."

Maybe removing the labels would remove the problem. The child reaches for solutions, hoping to bypass prejudice by inverting its language.

"I think that we could be much more tolerant," the child's voice pipes back again, "and that children should be tolerant even if their parents aren't."

Break the generational transmission of prejudice. Avoid imitation of negative parental influences.

"Why do you think many people don't like _____ people?"

"Some people think the _____ people shouldn't come into America," the child recalls sorrowfully. The ten-year-old's eyebrows crumple as their voice climbs to a crescendo. Something revs up, an anger, a frustration, a logic getting louder and gaining speed.

"People who think that should be eliminated themselves."

Oh no, was the child's tolerant reasoning slipping? *Falling into contradiction.*

"Have the people who say that go out of the country and only let them come back in when they don't say it any more."

Segregate the segregationists. The professor contemplates the child's penalty for prejudice. Remove the aggressors from the classroom—shut the door and bolt the gates until beliefs are smoothed out and homogenized. Erect borders to protect democratic sentiments from racist challenges. *Was it oxymoronic or brilliant? An instance of liberal or illiberal thinking? Rigid tribalism or simple immaturity?*

The first key to Else Frenkel-Brunswik's puzzle was that prejudiced children's rigidities were not constrained to one domain: they were everywhere. Rigidity spilled into every response, every reasoned thought and miscalculation. In interviews with children who had previously scored highly on prejudice in the questionnaires, the world was split into binaries. The world was black versus white, strong versus weak, good versus evil, friend versus foe, clean versus dirty, male versus female. "The absoluteness of each of these differences is considered natural and eternal," observed Frenkel-Brunswik, "excluding any possibility of individuals trespassing from the one side to the other." Clear borders between groups and categories had to be preserved.

When asked about gender roles, ethnocentric children claimed girls should not strive for careers or creative pursuits. In the words of one boy, "girls should learn only things that are useful around the house." An eleven-year-old girl observed that a "girl should act like a lady. Girls should not ask boys to date. It's not lady-like." Rigid rules were adhered to even when the rules implied restrictions on their own behavior, limiting possibilities and scope for movement.

The worst thing a woman could do, claimed one illiberal boy, was "to earn her own living—usually the man does that."

It is easy to brush these sentiments aside as products of the era—after all, the California teenagers of the 1940s had not yet grown into the advocates (or opponents) of the American women's liberation movement of the 1960s. But, even in the dark forties, when liberally oriented boys were asked about gender roles, they resisted binary thinking. When questioned about how girls ought to behave around boys, these liberal mini-men would say that girls should "talk about the things they like to talk about . . . the same as another boy would."

The worst thing a woman could do, argued one liberal boy, was "what she doesn't want to do."

For the liberal children speaking to Frenkel-Brunswik in the 1940s, the world of girls and the world of boys were overlapping circles, with more in common than not, rather than spheres barely touching, hovering side by side yet distinctly apart.

For the prejudiced children, on the other hand, all relationships were unequal and inadvertently abusive. Whereas liberal children viewed adult–child relationships as ones of reciprocity and warmth—relationships Frenkel-Brunswik called "mutuality"—the ethnocentric children perceived submission as the underlying logic of relationships. When asked about their teachers, prejudiced children would lament that "it would be better if teachers would be more strict" and "teachers should tell children what to do and not try to find out what the children want." The needs of children should be sidelined—argued these children—in favor of the pedagogical demands or private whims of their superiors. Many wished for strict male teachers and would accept a female teacher only if she was "*very* strict." The perfect teacher would be a perfect choreographer of discipline: she should "keep children organized" reported one child, "in the playground, in class, in lines."

Similarly, when probed about the ideal parent, prejudiced children venerated the ideal father as "strict: not soft on you." The model father is allergic to signs of affection. "When you ask for something," argued one twelve-year-old, "he ought not to give it to you right away." The father should be emotionally callous and unforgiving. "He spanks you when you are bad and doesn't give you too much money," reasoned another; "the punishment for talking back to parents should be a whipping." One girl even argued that a child "should be sent to a juvenile home for not paying attention to her parents." Dramatic penalties were proposed for minor transgressions.

Punishment was confused with justice, and justice was synonymized with pain.

These children asked—even begged and cried—for restrictions. They hoped for indifference, rationalized and at times glorified the violence committed against them. In the ways most antithetical to being a child—vulnerable, playful, periodically disobedient—these children were justifying their own oppressions, the militarization of their imagination and desires.

10

BRAINWASHING A BABY

How do you indoctrinate a child? Are elaborate tricks of mind control and propaganda necessary? How are ideologies learned and mastered? Is indoctrination always explicit or does the developing, maturing brain pick up on associations implicitly, subtly, illicitly, and later regurgitate them onward? And how early can indoctrination begin?

People who grow up in strict and devout ideological settings—whether religious or revolutionary or militantly nationalistic—can be more similar than they are different. Whatever the ideology, passionate ideologues exhibit striking parallels in reverence for authority, conformity, suppression of doubt or individuality. In an ideological community, life is viewed not as a series of beautiful "experiments of living," as the philosopher John Stuart Mill suggested, but rather as regimented *protocols* for living. There are rigid rules and all-important outward appearances: both must be perfected and sustained. Any hint of deviation or rebellion is dealt with swiftly, and sometimes violently. Aggression—physical, verbal, or atmospheric—is a common tool to evince power. Dogmatism and intolerance are the signatures of an ideological environment, even if they hide behind a sweet veneer of moral virtue.

Political scientists have long noticed that parents and their offspring are likely to share ideological beliefs. Statistically, secular and egalitarian parents nurture secular and egalitarian children, and conservative and religious parents nurture conservative and religious children. This is intuitive and unsurprising. The family is the ground for both biological reproduction and social reproduction. Genes, rituals, and worldviews are repeated, replicated, and recycled by each generation.

Yet the inheritance of beliefs is more nuanced than mere mimicry. Transmitting an ideology from mind to mind is not a cloning pipeline, stamping a personality or a political orientation onto the child. Neither is it akin to washing a brain—gentle soap and a sprinkle of oppressive bubbles and—*bang*—a fanatical infant. Luckily, it is not quite that easy. Becoming versed in an ideology is an engaged process of dynamic learning as well as selection. Inheritance is often partial and imperfect; the inherited can be revised and rejected.

To grow up in an ideological environment is to be trained in particular habits. Habits of attention. Habits of attachment. Habits of action. Habits that become "second nature," automatic and uncontested, guiding our gaze and correcting our tongue. Habits emerge from repetition. From the close and recurrent links created between a cue and a response. Every time a familiar trigger comes into view, there is a distinct action that must be executed. Every time a recognized context envelops us, we are—like a reflex personified—prompted to perform. On cue.

There is a relentlessness, a meticulousness, to the practice of habits.

Under habit, the brain forges tight connections between a cue and an action, until the mere presence of the cue—without careful deliberation—activates the action. Leave the house = grab the key. Meet a new acquaintance = shake hands or respectfully bow the head. Prepare for a religious ceremony = wash and purify the body in ablution.

Enter a place of worship = trace a cross across the body or brush lips against a spiritual symbol. Encounter the trigger = execute the learned sequence.

Under habit, an invisible hand grasps our jaw and directs our eyes in specific directions, so that what we see and what we notice are supervised. So that we learn to ignore certain objects or sensations or contradictions. We learn to enact actions on demand, the associations becoming so ingrained that we almost (almost!) cannot do otherwise, we cannot *not* perform the habit.

Habits are shortcuts that persevere even when the goal has vanished from view. A dog's habit surfaces when it performs a learned action even in the absence of a treat or a threat. *Sit. Roll over!* Human habits operate in the same way: habits lead us to complete an action for its own sake, even when the reward is gone. Even when we do not plan to or want to. Even to our own detriment. When we override our goals and desires, when we perform an action with a tinge of disgust, feeling pangs of regret and compulsion and inexplicable relief, but doing it anyway, we have encountered our own habits.

By definition, habits are always performed in the same way, the movements and sequences ritualized to perfection. In the spiritual habit as in the military habit, the believer moves precisely—*palms clasped together, feet perfectly aligned, the neck prepared to bow, eyelids closing and opening on command, agency is relinquished.* In the patriarchal habit as in the sporting habit, the believer disregards pain or discomfort—*evident by the scars of the blistered foot pressed into the high-heeled shoe or the perpetual wounds of the fanatical athlete.* Rituals performed in the service of an ideology are viewed as righteous regardless of how hurtful or ridiculous they may be.

Habits are about unquestioning replication and constancy. Yet at the same time, every performance of a habit strengthens its power, strengthens our conviction, strengthens the connection between a trigger and an action. Habits are never truly stable. Through repetition, habits can continually gain power, becoming more extreme. Performing a habit again and again has consequences. Our bodies are changed by the experience. A habit performed five times is, psychologically, different from the identical habit performed one hundred times. The more we repeat a habit, the more rigid it becomes. The more passionately we repeat a ritual, the more radical we become.

A brain engaged in a habit looks different from a brain acting in pursuit of a goal. When a habit inhabits the brain, the neural circuitry governing our behavior shifts away from the deliberative organs behind our foreheads in the prefrontal cortex and toward deeper and older structures in the brain, in the striatum that sits in the middle of the skull, at the center of us. The pattern of neurons firing in response to a practiced habit is different from the pattern of neuronal activity when we are making a deliberate goal-oriented action.

Neuroscientists can train mice and rats and pigeons and monkeys to act habitually, to perform actions even when the goal is absent, when the incentive is no longer there. Animals can learn to perseverate—that is, to persevere even once the reward is gone. Even when there are penalties for performing it, a truly deep-rooted habit will be replayed anyway.

Ideological groups and cults will purposefully instill habits and watch whether the believer upholds them in moments of suffering. The true believer will tolerate punishment to maintain a habit. *They may even relish the pain.* This is a sign that indoctrination runs deep; the believer will pay anything, may even pay with their lives, to sustain the devotion.

The most persistent habits, resilient to inconvenience or penalties, are ultimately addictions. This is why neuroscientists can train rats to become cocaine addicts—they teach them to enact a sequence of behaviors and reward them with cocaine until the sequence becomes a habit, until it is no longer a goal-directed behavior. Even if the rat endures an electric shock every time she pushes the lever that once delivered cocaine (but no longer does), she will continue to press. Frazzled and electrocuted but unable to stop, unable to shake the habit even when the reward no longer materializes and only punishment persists. Addicted, the rat repeats the habit even at the risk of torment, to the verge of death.

A strong, incorrigible habit is a form of an extreme devotion. A devotion to the promise of relief from some discomfort—a dissatisfaction, a hunger, a need for closure or certainty, an anxiety about the self. (A promise that in fact does not always provide respite from the frustration that drives us to rely on reassuring habits in the first place.)

To grow up in an ideological environment is to be trained in the rituals of devotion. Although devotion sounds lovely and peaceful, big-eyed and pious, devotion is demanding. Devoted, we de-individuate and become indistinguishable. Simultaneous kneeling in prayer. Boots marching with soldierlike precision. Chanting and moving together rhythmically. Devoted, we are submerged, anonymous. Letting go of the boundaries that separate self and other, we converge with others. Devoted, we are addicted to a promise, even at what should be intolerable costs to our well-being and bodily integrity.

Many people find synchrony—the demand of devotion—intensely beautiful and pleasurable. Training our attention to look outward rather than inward can feel meditative. Collaborative. The French philosopher Simone Weil thought it was freeing. "Attention consists of suspending our thought, leaving it detached," she wrote, "empty and ready to be penetrated." Weil's attention is a form of self-emptying. A prayer. Weil thought that such self-erasure was the epitome of grace.

But rituals of synchrony can also be dangerous, rendering individuals receptive to an array of oppressive forces. Disciplining our bodies and our attention outward, social rituals direct us toward others' needs, judgments, and appearances. The outer becomes more important than the inner. Doubts, questions, and inner sensations are silenced, and so the impulse to revolt is silenced too. Resistance cannot break through to the surface. We look away from ourselves.

The imagery of ideologies often invokes eyes—surveillance eyes, eyes that surround, eyes that censor, condemn, and control. At times benevolent but mostly intrusive. In fiction, myth, and reality, these omniscient eyes leave no action or thought hidden. George Orwell's Big Brother. Jeremy Bentham's panopticon prison. Satellite surveillance from above or algorithmic surveillance in our pockets. The Eye of Providence—with its all-seeing iris encapsulated in a pyramid, its lashes radiating into incandescent sun rays—features in the emblems of Christian temples, Jewish cemeteries, and Vietnamese Caodaist iconography. This eye graces the back of the United States' one-dollar bill and hides in the symbols of the cryptic Freemasons and Illuminati conspiracies. Power rests with the one who beholds all things and can change (or manipulate) behavior simply by being a persistent and moralizing witness.

Yet it is interesting to consider what happens when we invert the image (and hence the question). Rather than asking about the effect of feeling watched, eyes upon us at all times, listening in, we can think about how ideologies change our own sight and capacity for insight, *our own eyes and our own I's*. Ideological systems do not merely perceive us, giving us the chilling discomfort of having a perpetual spectator; ideological systems fundamentally refocus the spotlight of our perception, schooling our sensations and the scope of our visual attention.

This notion that ideologies might structure our consciousness, even our nonpolitical consciousness, was implicit in Else Frenkel-Brunswik's final studies, conducted throughout the late 1940s and early 1950s.

After her interviews with hundreds of children, each displaying varying degrees of intolerance, she began to wonder whether she could capture the prejudiced children's tendencies to split the world into binaries without relying on their tales and reports. Maybe testing sensory perception directly could foreshadow a child's prejudice. Perhaps the interview methodology itself could be abandoned and replaced with cognitive assessments of how children parse visual scenes and non-social scenarios, such as optical illusions or multistep problem-solving challenges.

To this end, Frenkel-Brunswik designed a series of "tentative experiments" to quantitatively assess whether the reverberations of ideologies could be felt beyond the realm of social beliefs. She and her colleagues invited a subset of the interviewed children back to the lab to test their cognition and memory.

The researchers gave participants mathematical exercises that demanded long and complex methods to arrive at the solution and later presented the participants with exercises that could in fact be solved using a simpler approach. Prejudiced children perseverated with the old, convoluted method, while the unprejudiced children flexibly reevaluated the arithmetic problems and solved them quickly and efficiently using the new approach. The level of rigidity children displayed on this arithmetic task was correlated with the rigidities they exhibited in the interviews about moral values, ethnic minorities, and gender roles.

Similar parallels were found between social prejudices and mental rigidity when the children were asked to trace their way flexibly out of a labyrinthine maze on a map or to complete memory tests in which they listened to stories and had to recall details afterward.

In one memory test, a story was recounted in which the characters were all pupils of a school to which new students had just been admitted. Each pupil was described in terms of a personal accomplishment, such as their sporting successes or musical achievements, and a background fact, such as information about the pupil's economic status or race. After listening to the anecdotes, the participants were asked to write a summary of their recollections and opinions about the characters—who was good and who, if anyone, was villainous. "We may conceive of the story as a piece of reality," Frenkel-Brunswik explained, "and ask ourselves what changes this reality undergoes in the memory of the children, especially in the direction of an elimination of ambiguities and other complexities." What details would children remember? How accurately could they paraphrase the story?

Liberal children tended to recall more accurately the ratio of desirable and undesirable traits—their memories possessed greater fidelity to the story as it was originally told relative to their illiberal peers. In contrast, children who had scored highly on prejudice had "negativistic tendencies in the distortion of the story content," highlighting or inventing undesirable traits for the characters from ethnic minority backgrounds. "In the recall of the prejudiced children the story gets generally more simplified and less diverse," Frenkel-Brunswik reported.

A close analysis of the participants' recollections uncovered that the most prejudiced children tended to "stray from the content of the story . . . telling stories which show almost no relation to the material presented," and at the same time they tended to parrot "certain single phrases and details" from the original material. Prejudiced children swayed between neglecting the stimulus in front of them—fabricating fictions instead—and clinging so closely to the stimulus that they could perfectly and eerily mimic the storyteller. Interpreting this combination of disintegration and rigidity, Frenkel-Brunswik suggested that "both these patterns help avoidance of uncertainty, one of them by fixation to, the other by tearing loose from, the given realities."

Next, Else Frenkel-Brunswik turned to perception. She wanted to study how prejudiced and unprejudiced children evaluated visual spectrums of colors or images of cats gradually morphing into dogs in successive panels of illustrations. Who would accept ambiguous mixtures and who would latch on to first impressions and refuse to change their interpretations? Perhaps looking into a child's eye would be enough to detect the thick gloss of rigidity, the potential for fundamentalism.

The professor places a card on the interview table.

"Please indicate this card's color."

"Red!"

The child beams with sly self-satisfaction. Easy.

"And what about the color of *this* card?"

Lifting a square panel coated in a solid magenta, the professor remains expressionless and nudges the card closer to the child's face.

"Also red!"

"And the next card?" A bright violet this time.

The child's streak is slowing, and now a pause precedes the response.

"It's . . . also red."

"And this card?" A jammy plum.

"Maybe that's purple?" Uncertain now, the question mark reverberates.

The professor continues with stamina, raising another panel. "What color is this card?"

A muted indigo stands between the interviewee and the professor.

"Purple."

Shuffling the cards, the professor jumps to the last card in the deck and holds it up in front of her face. The professor's head disappears behind a sky-blue square. She wants to probe how easily a child can move from one color to the next, how rapidly they can shift color categories and how willingly they accept ambiguities and changing hues. When does a red stop being red and become purple? Some individuals adapt swiftly. For others, confessing that a color no longer matches the original category seems a travesty. Admitting change is like admitting defeat.

"I guess that's not purple any more." A blink. "It's definitely blue."

The pattern was there. "Preliminary." But the statistics backed up her hypothesis. The most prejudiced children struggled with the shifting colors, preferring clear classifications and pretending that the ambiguities were invisible and unimportant. The liberal children were sensitive to the changing hues—they easily disengaged from their first sensory judgment and moved on—accepting that colors are spectrums of tones and tints, rather than neat binary categories. Even in the domain of pure colors, liberal children felt comfortable with ambiguity, tolerating the transitions between the bands of the rainbow, allowing pigments to blend into each other. Dyes do not need to be shoved into distinct grades: they can coexist in happy ambiguity.

Rigidity in one domain leaked into rigidities elsewhere. The segregating mind divided everything up. Tendencies to socially segregate minorities from majorities were mirrored by a tendency to segregate colors into binaries as well. A person who chops up the perceptual world is also a person who brutally cuts up the political world.

For Frenkel-Brunswik, the ways in which children resolved perceptual ambiguity revealed how they resolved interpersonal and political ambiguity. "It is as if any stimulus—or what seems to be 'the' stimulus in the person's interpretation—is playing the role of an authority to which the subject feels compelled to submit," reported Else Frenkel-Brunswik on the prejudiced participants' behavior. "Situations which seem to be lacking in firmness are apparently as strange, bewildering, and disturbing to the prejudiced as would be a leader lacking in absolute determination. With internal conflict being as disturbing as it is in this group, there apparently develops a tendency to deny external ambiguity as long as such denial can be maintained."

"Tentative experiments," wrote Frenkel-Brunswik excitedly, tempering herself with caution—the participant pool was small and the theory was not yet fully fleshed out. The data were not ripe for big and bold conclusions. Precision and rigor were of utmost importance. But so were these promising discoveries. However "scattered" and "preliminary" her conclusions, she had to tell the world about what happens to the prejudiced mind. Why totalitarianism tutors not only our conscious beliefs but also our unconscious patterns.

Tragically, these nascent experiments never matured. Just as Else was beginning to receive international recognition for her innovative approach, her husband Egon—after battling a long and painful illness—committed suicide. This misfortune halted Frenkel-Brunswik's scientific productivity and soon afterward, in 1958, she put an end to her life too. Her ideas lay largely dormant. Remnants were misappropriated or misattributed to male collaborators and students.

Before she passed, the spirit of Frenkel-Brunswik's project was turned into the celebrated book *The Authoritarian Personality*, published in 1951. Frenkel-Brunswik was named the book's second author, preceded by the famous political theorist Theodor Adorno. But looking closely at the ideas and the methods, it is clear that they are rooted in Frenkel-Brunswik's individual differences approach. Theodor Adorno had never led a serious quantitative study before the publication of this book. In fact, Adorno was suspicious of logical positivism, the Viennese school of thought that championed scientific experimentation as the means to create and verify knowledge. Else, on the other hand, was deeply embedded in the Viennese circles of the logical positivists, more than a decade before *The Authoritarian Personality* was ever conceived.

Frenkel-Brunswik's entire career was dedicated to the study of personality, while Adorno had barely mentioned it in any of his writings. It is therefore unclear why Frenkel-Brunswik did not receive more primary credit for a project that she had been working toward for decades, and which none of the coauthors of *The Authoritarian Personality* continued to pursue afterward.

Although Frenkel-Brunswik's contributions have been historically sidelined and underacknowledged, she left a trail for future researchers to follow. In 2015, sixty years after Frenkel-Brunswik's death, I discovered her work, finding striking resonances with my own investigations of the cognitive characteristics of the ideological mind. Although our research was conducted more than half a century apart, in dramatically different political climates, there was nonetheless a convergence in our interests in decoding the unconscious markers of dogmatic thinking. In the new millenium, I built upon Frenkel-Brunswik's pioneering insights using the rigorous methodologies of the cognitive and brain sciences, just as a new wave of ethnocentrism and xenophobia was becoming apparent in the Europe she left behind.

11

THE RIGID MIND

Please press ENTER when you are ready.

Welcome back! I hope you are enjoying these games. For your next challenge, you will see an image of an object on the screen, such as a paper clip. You will have two minutes to come up with as many possible uses for that object. The possible uses can be conventional or unconventional, mundane or avant-garde.

For example, if you are presented with a picture of a paper clip, your potential ideas could include: using the paper clip for holding bunches of paper together; using it as a hairpin; using it to reset your smartphone; using it as an uncomfortable toothpick; using it to pick a stubborn lock; hooking it together with other paper clips to form a decorative bracelet; engineering a mini sundial; bending paper clips and connecting them together to form a small ladder for a mouse or an unusually slim hamster.

Your ideas can be as extravagant or as straightforward as your imagination allows. As long as your solution is functionally feasible or theoretically workable, it is a worthwhile idea.

Now it's your turn. Ready?

For the next two minutes, give as many possible uses as you can for a . . . BRICK.

You write:
Build a house.
Build a chimney.
Build a school.
Build a prison with a watchtower.
Build a castle for a captive princess.
Smash a window.
Use as a plant pot.
A door stop.
A paper weight.
Grind coffee beans.
Feed the brick to a brick-eating monster.
Crush a large spider.
BEEP!
Time is up.

You might feel accomplished or inadequate. A little surprised by the paths your mind took to come up with these solutions. You might wonder what ideas you missed or what your mind could have generated with just a little more time and a little less outside interference.

This game is called the Alternative Uses Test. It is a classic measure of creativity, assessing the degree to which you can produce a wide range of creative ideas. Whereas the Wisconsin Card Sorting Test taps into *reactive* flexibility—how well you adapt when the external world changes—the Alternative Uses Test measures your *generative* flexibility. Generative flexibility reflects how malleably you can think internally, running through numerous ideas and visualized scenarios, fabricated spontaneously, with few external stimuli. The Alternative Uses Test quantifies how widely you search your semantic networks of concepts and associations, how quickly and malleably you rummage through solutions and possibilities; the degree to which you can think outside the box.

For some, a flood of ideas flows out of fingertips and onto the screen. For others, the generation process is slower and less prolific, more like a light shower than a biblical monsoon.

In order to transform text-based qualitative responses, which are sometimes random or rambling, into quantitative data, cognitive scientists begin by counting the number of reasonable ideas provided within the time limit. This indexes the participant's *fluency*. Fluency captures how quickly and easily you produce viable ideas. Fluency measures how swiftly you search through your memory and conceptual networks to produce a multitude of relevant and appropriate solutions. Many ideas = fluency.

Scientists also measure your *elaboration* capacities, seeing how intricately detailed your ideas are. Do you tend to give highly vivid and detailed answers—"build a castle for a captive princess"—or do you stick to the basics—"build something"? Lucid, descriptive, and distinctly expressive ideas = elaboration.

Researchers may try to capture your *originality*, estimating the degree to which your ideas are unusual and rare. If you tend to think outside the box, if you come up with ideas that very few people could envision, then you are an original thinker. If, on the other hand, you mostly think of building houses and fences, you are operating very much within the box. Uncommon and unconventional ideas = originality.

Finally, and most interestingly to me, it is possible to quantify your *flexibility*. Flexibility is measured by how many different kinds of uses you were able to generate for the object at hand. A highly flexible person would provide a wide range of solutions with diverse functions; they might suggest that a brick could be used for purposes of construction or weaponry or gardening or the satiation of imaginary brick-eating monsters. A more inflexible thinker would provide only a narrow set of ideas. For instance, they might offer many things to *build* with bricks—houses, castles, schools—but wouldn't imagine other possible functions, such as using a brick as a door stop or a kitchen utensil. The inflexible person struggles to see the plurality of functions a single object can possess. To them, things have fixed and singular essences. On the other hand, for the flexible thinker, nothing is locked to its most common role. The ability to imagine diverse functions = flexibility.

After inviting thousands of participants and recruiting an army of eager research assistants to analyze each individual's data, I found that mental flexibility on this creativity task is uniquely linked to an opposition to dogmatism—to intellectual humility. The more flexibly people performed on several iterations of this Alternative Uses Test, the more responsive they were to alternative viewpoints and the more willing they were to update their beliefs in light of reliable evidence. The more cognitively rigid people were, the more ideologically rigid they were. This correlation was true regardless of age, gender, or educational attainment. The pattern held for the young and the old, the highly educated and the less, regardless of demographic characteristics.

Notably, out of the multiple metrics we can extract from the Alternative Uses Test, flexibility was the only significant predictor of dogmatism—people's fluency, elaboration, and originality were not significant predictors of intellectual humility. There is something special about flexibility as a style of thinking. Even when we statistically control for intelligence, cognitive flexibility remains a significant predictor of intellectual humility. Even for the least intelligent among us, mental flexibility can ensure that we are receptive to evidence and alterity.

Why is the adaptability of our minds—our reactive and generative flexibilities—so important?

Because to behold a box or a brick or a metallic paper clip and spot opportunities for creativity is to overcome immediate appearances. To dismantle habits. To break fixed essences in order to reveal their instability, their fluidity. To transmute the invisible, the inconceivable, the unobservable or unobserved into something visible, possible, and new. When we are flexible, we transition easily; unencumbered by static *beings*, we focus on ever-moving *becomings*.

Our creative imagination is linked to our ideological imagination. A person with a flexible imagination is more likely to be flexible in evaluating ideological claims. Rigid imagination, rigid ideologies. "What art and morality have in common is creation and invention," suggested the French philosopher Jean-Paul Sartre in his celebrated 1945 lecture *Existentialism Is a Humanism*. "Has anyone ever blamed an artist for not following rules of painting established *a priori*? Has anyone ever told an artist what sort of picture he should paint?" The cognitive capacity to break away from the norm underpins the artist's avant-garde experiments as well as the nonideological thinker's shunning of dogma and orthodoxy.

Our thoughts are clustered by associations. Concepts or objects that co-occur frequently become strongly connected, such that activating an image of a brick most commonly leads to images of joining bricks together into buildings, houses, defenses. For our thoughts to move to an entirely different dimension, another plane, and think of alternative uses for a brick—crumble it up, step on it, use it to crush garlic or grind stubborn peppercorns—we need to truly switch mental tracks, modes of thinking; let our associations travel far and wide; hold the object in our mind, rotate it and let it metamorphose into different shapes, textures, contours, and configurations. That is the beauty of a task like the Alternative Uses Task: it captures our capacity to alternate, to deviate, to mutate an item into something entirely other.

This capacity is not valued or cherished by all. There is a sinister and racist history to research on the links between perceptual flexibility and ideology. In the 1930s, the Nazi psychologist Erich Rudolf Jaensch created a personality typology consisting of contrasting types of perception—some people were endowed with the qualities that would make the ideal German Nazi, and other people possessed the characteristics that would render a person a threat to the Nazi regime. For Jaensch, the "J-type" was a person whose perceptual experiences were unambiguous, firm, definite, machine-like, and "integrated"—their associations clustered together in predicable ways. The J-type was rigid, masculine, aggressive, and a reliable Nazi Party member of Aryan ancestry. In contrast, Jaensch posited the existence of an S-type, a synaesthesia-prone individual whose sensory tendencies were looser, more fragmented, more infected by emotions, and "liberal" in all the ways that Jaensch and the Nazis detested. The S-type was likely to be eccentric and individualist. According to Jaensch, the S-type was likely to have Jewish or racially mixed ancestry. Jaensch claimed S-types were probably communist or "Parisian" or of Asian descent. Psychological flexibility was a red flag. Narrowness of perception was a virtue.

Jaensch's science was not only abhorrently racist but also methodologically flawed (as most racist research tends to be). Else Frenkel-Brunswik observed that "the sampling technique and the statistical significance and validity of the works of Jaensch is of the shoddiest kind and cannot stand under scrutiny even of the mildest kind." Yet it serves as a crucial reminder that scientific investigation is not devoid of moral values. Whether we value a flexible mind or a rigid mind—a mind tolerant of ambiguities or a mind seeking to eliminate them—is a reflection of whether we desire a versatile world or an unchanging homogenous one.

Rigidity erects mental borders, high walls over which we struggle to climb. Sometimes these mental borders mobilize us to build real geographical borders too. Whether we can make abstract leaps matters because our inflexibility shapes our concrete political beliefs and voting behavior. Across a series of studies with British and American participants, I found that cognitive rigidity influences not only our inner world but our external political realities too.

If we switched roles for a moment, and you sat me down on the gray chair and told me that you have one simple question for me—the question of where I am from—I would quickly feel a wave of frustration surge. I would purse my lips and crunch my nose, squeezing my cheeks in an act of containment—so that nothing accidentally slips out—and maybe I would avert my gaze and stretch my neck or shake my head in discomfort. My body will refuse the question. I will not have an answer for you. I am not from anywhere. The story of my identity extends further than a one-line, two-syllable answer. This is true for many of us living in mobile and multicultural societies. I own no flag and my accent betrays little coherent ancestry. I live in a different place from where I was born, and I was born in a different country from where my parents grew up, which is again different from the now-forgotten lands where my grandparents took their initial steps and tasted their first words. My grandparents grew up speaking one language, and my parents cultivated the slang of another, while I learned the grammars and nuances of entirely different scripts. We all came of age on different continents, hearts breaking under different skies, overlooking different seas, our secrets and curses whispered in different tongues.

I am a "citizen of the world."

However, for many people, a citizen of the world is a "citizen of nowhere"—as expressed by the former UK prime minister Theresa May, one of the chief administrators of the UK's exit from the European Union.

In 2016's Brexit referendum, and at other times past and future, politicians asked voters how they would like to redraw borders. This question was intensely personal. It was a question of how a private identity is externalized, how rigidities become carved into the geopolitical atlas.

Brexit was unprecedented because it did not map onto traditional political parties. The referendum was a unique instance when party affiliation or elite cues did not clearly instruct people how to cast their votes. The UK's primary political parties offered little consistent guidance. Each person needed to decide their opinion on their own, with few anchors to rely on for their decision. There were few ready-made ideological paths to follow. This makes Brexit a perfect opportunity to study individuals' authentic ideological positions—or at least how they independently process the ideological debates that surround them, without leaning on past identities for support.

Brexit did not embody the caution of conservatism, its hesitancy. Brexit upended the status quo and required a dramatic shift. But at the same time it was a nostalgic return to an imagined past, a retreat from mobility and multiculturalism. As a result, the UK's exit from the EU was a movement of both regression and change.

This change did not resemble the revolutionary spirit of progress sometimes beloved by the political left. And yet some leftists viewed it as an opportunity for rebellion—a revolt against the EU's bureaucracy, inefficiency, and excessive centralization of control away from ordinary citizens. With many circulating interpretations of the meaning and consequence of the same political event, Brexit created a whole new axis for British politics.

Among the swirls of information and misinformation, people were left unescorted to distinguish between truths and falsehoods, bent statistics and fraudulent claims. Each was forced to decide whether they felt an affinity with the notion of well-defined borders—*Sovereignty!*

Control! Britishness!—or with ill-defined boundaries that could be freely crossed and toyed with. *Who wants to play?*

A closed Britain was a greater Britain, many leaders exclaimed, and its two-syllabled mantra—"Vote Leave"—was fun, rebellious, and sexy. It overshadowed the clunky, bespectacled "Britain Stronger in Europe." "Vote Leave" was almost onomatopoeic—a symmetrical alliteration of the hard fricative sound, forcing the speaker's top teeth to lightly bite the bottom lip with every utterance. There is a sense of severance, of self-injury, of breathless withdrawal.

Nationalism is most commonly correlated with right-wing politics rather than left-wing politics. Yet this is not inevitably or universally true. Nationalistic attachment is, in principle, distinct from political ideology. Whereas nationalism tends to focus primarily on perceptions of national superiority, on idealization of the nation and its dominance or history, many diverse issues are clustered into party politics. The conservative-versus-liberal axis covers views as eclectic and unrelated as economic policies, education, environmental protection, the primacy of religion, and civil rights. The particular clustering of policies that a country will find across its political parties depends on its cultural history, electoral plurality, and the economic inequalities that plague its systems.

Nationalism is cleaner. The mythology of every nation pivots around fights for independence. A need to separate an *us* from a *them*. "What is love of one's country?" ponders a character in the novel *The Left Hand of Darkness* by the American writer Ursula K. Le Guin. "Is it hate of one's uncountry?"

At the heart of nationalistic ideologies are strict categories and rules for what is or is not part of the nation or national culture. Following the UK's vote to leave the EU in 2016, I saw an opportunity to study whether the strictness of national identity was linked to cognitive rigidity. I hypothesized that nationalistic thinking may be an instance of a general tendency to categorize information rigidly and to process information in an inflexible manner, such that individual differences in cognitive inflexibility would be predictive of support for Brexit in the United Kingdom's EU referendum.

Studying hundreds of British participants surveyed after the referendum, I found that individuals' cognitive rigidity predicted their nationalistic beliefs in the context of Brexit, such that more cognitively rigid individuals were more likely to vote to leave the EU. Cognitively rigid participants were more likely to agree with statements such as "a citizen of the world is a citizen of nowhere." They were also more likely to believe that national citizenship is an inflexible category, and that strong borders need to be erected between nation states. *Let nothing in.*

In the experiment, I found that the degree to which the brain adapts to change and how rigidly it constructs internal boundaries between conceptual representations was linked to the individual's desire for external boundaries to be imposed on national entities and for greater homogeneity in their cultural environment. A person's cognitive style echoed their ideological style.

Mental rigidity was also related to how rigidly people interpreted the task of governance. When asked to indicate how strongly they agreed with the statement that "the UK government has a right to remain in the EU if the costs [of leaving] are too high," the cognitively rigid participants believed in no compromise—the UK government must do all that is necessary to leave the EU. From the perspective of the rigid thinker, the government ought to be paralyzed by the referendum, even if it means its responsibilities for efficient, ethical, and resourceful governance are abdicated. No cost would be too high.

The relationships between cognitive rigidity on behavioral tasks such as the Wisconsin Card Sorting Test and ideological rigidity in the context of Brexit all persisted after statistically accounting for age, gender, and educational attainment. This illustrates that information-processing styles in relation to perceptual and linguistic information may also be drawn upon when dealing with political information. The rigid mind is inflexible in sensory games, in acts of imagination, and in its unrelenting rigid convictions about right and wrong.

This raises questions about the rigid mind's affinities toward the political right and left. Is it possible that the diehard Democrat and the unflinching Republican are more cognitively similar than they are different?

Since political psychology's earliest days, there has been an assumption that the political right is naturally rigid. Right-wing ideologies tend to fortify traditions and hierarchies and resist change, and so the minds of right-wing ideologues are more likely to be rigid—not only in matters of politics but also in general. Else Frenkel-Brunswik's work with Theodor Adorno on *The Authoritarian Personality* insinuated that inflexibility was uniquely a phenomenon that gave rise to conservative impulses and right-wing xenophobia. Social and personality psychologists flocked to the task of proving this rigidity-of-the-right effect by testing individual differences in rigidity with the tool they loved best: self-report questionnaires.

But this was an optimistically lenient approach. The problem with self-report questionnaires is that people can lack self-insight, believing they are more flexible or rigid than they truly are. People can also modify their responses according to what they perceive is most socially desirable. In right-wing communities that uphold values of stability and stasis, a right-wing believer may not be incentivized to honestly report their flexibility. In left-wing communities that advocate for change and disrupting conventions, a left-wing believer may not like to see themselves as inflexible and routine loving. Asking participants to assess their own cognitive rigidity and flexibility is therefore likely to elicit answers that reinforce existing social representations rather than tap into genuine cognitive styles. This is why cognitive and behavioral approaches that measure cognitive traits in objective and unconscious ways are paramount for generating reliable conclusions.

A similar problem materializes when measuring a person's ideology: their self-reported affiliations are confounded with reporting biases and limitations. Self-identifying with a political "side" tells us little about a person's policy preferences—if they even have clear policy preferences to begin with—or their degree of ideological extremity. This tangle of measurement has led some thinkers to postulate that we can never measure an individual's ideology. The Italian Marxist thinker Antonio Gramsci ridiculed the idea that ideologies are anything but collective and unconscious. In his *Prison Notebooks*, Gramsci scoffed: "Obviously, it is impossible to have 'statistics' on ways of thinking and on single individual opinions." More reflectively, the political scientist Philip Converse conveyed that "belief systems have never surrendered easily to empirical study or quantification. Indeed, they have often served as primary exhibits for the doctrine that what is important to study cannot be measured and that what can be measured is not important to study."

Measuring an individual's liberalism poses challenges for the scientist trying to capture it and make it surrender to empirical analysis. When a person defines themselves as liberal, there are two common interpretations of the label. The first is as a form of open-mindedness, a scientific orientation, a receptivity to evidence, a respect for plurality and individual liberties. This is liberalism as an antipathy to ideology in all its forms. The second interpretation equates liberalism with left-wing worldviews, which are traditionally concerned with individual liberties and diversity but also entail explicit policies regarding the distribution of resources, sympathy for the marginalized and disadvantaged, and the role of the state in providing support to its citizens and organizing social life in egalitarian ways.

The specifics of this left-wing liberalism vary country to country, party to party, decade by decade. This is visible in Latin America, where many governments have embraced left-wing politics, but each has adopted a strikingly different policy position on the degree to which public services and industries should be nationalized or opened to free-market forces. "Leftist" causes may be taken up by "rightist" parties in a different election year or under the auspices of a different mood or language. Whereas "leftist" parties often seek to increase public spending and "rightist" parties promote austerity, in many formerly communist European countries the pattern was reversed between 1989 and 2004 as they transitioned from socialism to democracy: left-leaning parties tended to prefer policies that reduced governmental spending on health and education whereas right-leaning parties tended to capitulate to voters' desires for greater social provision and state intervention. The forces that thrust voters toward certain policies and party manifestos and not others are highly contingent on time and place and body.

In fact, the psychological traits that prompt a person to adopt right-wing views in Western democracies have been shown to render people more prone to adopting left-wing views in formerly communist countries. Context matters. Status quos shift. The link between personality and politics depends on circumstance. A personality marked by a need for structure and routine tends to lead people toward right-wing traditions in the United States and Europe. Yet the same trait can render left-wing collectivist ideologies more alluring in places where authoritarian left-wing regimes once flourished, such as in former Soviet republics or Latin American countries.

Many liberals embody both meanings: open-minded, science-minded, and at the same time ardently supporting left-wing causes. But there are also self-identified liberals who belong primarily to one flavor. There are illiberal leftists who are dogmatic, intolerant, and willing to commit violence in the name of their cause. They can become leftist extremists: passionately obsessed by their rigid doctrine and rigid identity. But if we define liberalism as openness to evidence and debate, it is oxymoronic to label a person a "liberal extremist"—pluralistic evidence-based thinking is antithetical to extremist thinking.

The problem of definition and measurement escalates when we consider that when psychologists try to measure a person's political ideology, the most typical method is to ask people to rate themselves on scales of "extremely liberal," "liberal," or "somewhat liberal," all the way to "somewhat conservative," "conservative," or "extremely conservative." But the way in which different readers will interpret the term "liberal" varies widely. A person committed to liberalism in the first sense may claim to be "extremely liberal" to demonstrate their strong commitment to pluralistic and evidence-based thinking and debate, and their opposition to rigid identity groups. Yet a dogmatic left-wing ideologue who exhibits severe intolerances may also self-identify as "extremely liberal." The "extremely liberal" bucket is thus filled with diverse and at times contradictory personalities. As a result, we need to be cautious in making sense of empirical research that compares self-defined conservatives and liberals; there are multiple axes of difference within and between these groups. Ideological substance and ideological style are confounded. When research relies on ideological labels, it can tell us more about the functioning of conservative ideologies than about liberal philosophies or left-leaning ideologies because different versions of tolerance and intolerance become muddled and intertwined in the "liberal" category.

In order to unknot these threads so that we might rigorously test the relationship between cognitive rigidity and partisan identities, innovations in measurement are necessary. One method I adapted is to measure a person's identity fusion. Participants are invited to draw a large circle and give it the name of their most beloved political group, or the political group that feels the least objectionable. A liberal American participant might call it the "Democratic Party." A few inches to the left of this big circle, participants are shown a little circle labeled "Me." Next, participants are asked to move the little "Me" circle toward the circle of the "Democratic Party" and place it in the position that feels right—the position that accurately captures the relation between their personal identity and the collective identity.

If it were you, what would you choose? Where would you place "Me" in relation to the big circle denoting your favored political party? Are the circles touching? Overlapping? Would you fully immerse the private circle in the larger public sphere? Would you be swallowed up entirely by the collective—your identities tightly fused?

With this visual measure of identity fusion, it is possible to gather a sense of how much you submerge your personal identity in the collective. Some of you are completely inside—no edges, fingers, or toes peeking outside the totality of the group sphere. Some of you overlap a little with the political identity—but not too much. Others may have dragged their little "Me" circle *away* from the group circle. Maximizing the distance. No overlap.

In a study with over 700 Americans in 2016, I asked participants to complete this identity fusion scale twice. Once for the Democratic Party and once for the Republican Party. I then subtracted the identity fusion of their least favored party from the identity fusion of their preferred party. I could then construct a spectrum from strong left-leaning partisanship to strong right-leaning partisanship, without asking people to self-identify via political labels or splitting them into two groups. The continuum began with extreme-left partisans, extended to moderates and independents, and ended with extreme-right partisans. A strong partisan's identity is exclusively fused to the political right or to the political left. For nonpartisans, the absolute identity fusion is closer to zero, with no overwhelming preference either way.

I then studied how participants performed on tests of mental flexibility. I found that those in the moderate middle performed best on neuropsychological tests of cognitive flexibility and those on the outskirts performed worst.

I mapped participants' cognitive flexibility along the axis of political partisanship, with each participant representing one data point, a dot on the graph. Looking at the dots of hundreds of participants, they formed an arc, a parabola, a rainbow shape. Instead of the presumed linear line that would suggest a rigidity-of-the-right, we see a rigidity-of-the-extremes. The most extreme leftists ranked low on cognitive flexibility relative to nonpartisans, who represented the peak of the arc, the most cognitively flexible. Moving from nonpartisans further to the right, cognitive flexibility dropped once more for extreme rightists. Those on the extreme right exhibited a diminished mental flexibility on a number of tasks measuring adaptability in visual perception and linguistic puzzles.

The extreme right and the extreme left were cognitively similar to each other. Both extremes struggled to adapt, to invent, to change mental schemas, even in neutral and nonpolitical situations. This rigidity-of-the-extremes effect harks back to old horseshoe theories of politics—the idea that the extreme left and extreme right are ultimately similar in their intolerance and rigidity, that the left–right continuum is bent rather than straight, that fascism and communism end up meeting at the edges.

From the perspective of the individual, these results suggest that heightened cognitive flexibility may be a protective barrier against all kinds of extremism, left or right. At first glance, the clear conclusion seems to be that the rigidity-of-the-extremes hypothesis prevails and the rigidity-of-the-right hypothesis is doomed to history. *Not so fast.* When we look closely at the rainbow-shaped curve we see that its peak is not aligned precisely to the center point of the left-to-right continuum. The peak is slightly off-center to the left, such that the most flexible individuals are nonpartisans who lean to the left. The rigidity-of-the-extremes may be compatible with, rather than antagonistic to, the rigidity-of-the-right hypothesis. The most psychologically inflexible people are extreme partisans, regardless of their allegiance. But the most flexible individuals are nonpartisans whose support tilts toward the left while resisting joining their identities too strongly with an established political party.

In further studies, I found that cognitive rigidity not only predicted ideological extremity in relation to social identities—it also accounts for extreme behavioral intentions. When we examine the most extreme of acts—the readiness to harm, to kill, and to be killed in the name of an ideological group—the results convey an even more unsettling message. A whisper of more sinister conclusions to come.

You are standing on a bridge above hissing train tracks.

A trolley has lost control and is speeding along the railway tracks, heading toward five fellow compatriots—members of your country or community. You see that without your intervention, the anonymous, innocent people will be killed by the merciless driverless trolley racing toward them.

You have two options. Do you let your five compatriots on the tracks die under the wheels of this fast-sprinting trolley? Or do you jump in front of the trolley, sacrificing your life and saving theirs?

This is a variant of the famous philosophical conundrum known as the trolley problem.

The clock is ticking, each passing second thrusting you closer to a cruel and deadly outcome. You must choose. Time is running out, racing ahead of you. What will you do? Sacrifice your life and save your fellow group members? Or preserve your life and sacrifice theirs? Will you choose life or death? Personal safety or ideological martyrdom?

It's over. You chose.

What did you choose?

Only you (and I) know.

And now I want to ask the following question: *How certain are you in your decision?*

On a scale of 0 percent to 100 percent certainty, how confident are you that you would actually behave as you indicated?

A thousand people answered this moral dilemma for me. A thousand different responses and certainties. What I discovered is that individuals prone to self-sacrifice—more willing to die for their group—are more cognitively rigid on tests of reactive and generative flexibility. Cognitive rigidity lends itself toward extreme martyrdom decisions, whereas cognitive flexibility protects against proclivities for self-sacrifice. And there is more. The more confident a person was in their decision to self-sacrifice, the more cognitively rigid they were too. By contrast, confidence in self-preservation was not related to inflexibility. The conviction to sacrifice oneself—which some call martyrdom and others classify as altruism—is specifically connected to rigidity.

With this moral dilemma, I can draw the contours of your personal and ideological sympathies, where the moral spheres of belonging begin and end, where the edges become blurry or come into focus. Who is worth dying for? What is sacred in our eyes? And why is it sacred?

Violence against the self in this context is ambiguous—it can indicate multiple and competing perspectives on the world. Does it reflect a utilitarian approach that sees five lives as inevitably more valuable than one? Or does it betray a belief that self-sacrifice is a glorious and courageous act that should be idolized?

Violence against others is more common and easier to parse than violence against the self. So another measure is to study the degree to which you support violence against outgroups that (appear to) threaten yours. Would you push an outgroup member if they mocked your ideological group? Would you hurt someone who insulted your nation, your religion, your neighborhood, your team? How quickly and confidently do you injure others to protect the honor of your ideological group?

In experimental studies, I found again that your cognitive rigidity is correlated with your endorsement of ideological violence against an outgroup. The more inflexible you are, the more willing you are to harm others in the name of your group. Our cognitive tendencies are stitched with our ideological proclivities.

The origins of our deepest convictions can be traced back to our cognitive styles and personalities—but these too have their own origin stories. Whether we are proud or ambivalent about our ancestry, the bodies we inhabit are embroidered with our genealogy. A history is concealed within our biology, channeling our present commitments and future possibilities.

12

THE DOGMATIC GENE

WHISPER: Is it genetic?

WHISPER BACK: Is what genetic?

WHISPER AGAIN: Rigidity. Is my level of rigidity determined by my genes?

WHISPER SOFTLY: Yes, partially.

[*Gasp.*]

ANOTHER WHISPER: What does that mean?

A SOFT WHISPER: It means that each person begins with a different degree of susceptibility to rigid thinking—and that this susceptibility is partially encoded in their genes.

WHISPER AGAIN [*anxiously*]: How do we know this?

A CONFIDENT WHISPER: Scientists have studied the neural circuitry of rigid behaviors—in animals, in addicted individuals, in psychiatric conditions marked by compulsive behaviors or repetitive thoughts, in the whole spectrum of functional flexibilities and atypical inflexibilities—and discovered a chemical that governs much of our ability to adapt and discover, switch and update.

WHISPER [*grows more excited, less anxious now*]: Have I heard of this chemical?

WHISPER BACK [*gladly*]: Of course you know it; it is called—

Many of us have heard of dopamine—the feel-good neurotransmitter—the chemical released when we experience reward, thrill, pleasure, or excitement. It is the chemical messenger that signals when to feel a high and when to plunge into a low. It alerts us when our expectations have been met or when there is a deviation, an error in our predictions. By guiding our reward system, dopamine can control what we like, what we hate, whom we despise, and when we are surprised. In short, dopamine directs our learning and our responses, our habits and our addictions.

I discovered that the most rigid individuals possess specific genes that affect how dopamine is distributed throughout the brain. In the largest study to date, looking at thousands of British participants, I found that the individuals who are most cognitively rigid have a genetic predisposition that concentrates less dopamine in their prefrontal cortex, the decision-making center of the brain, and more dopamine in their striatum, the midbrain structure that controls our rapid instincts. This is significant. If our psychological rigidities are grounded in biological vulnerabilities such as how our brains produce dopamine, it becomes possible to trace the pathways between our biology and our ideologies.

WHISPER: Is there a dogmatic gene?

ANOTHER WHISPER: And why do the clues lie in dopamine?

Dopamine is an organic chemical that—when drawn out into a three-dimensional molecular diagram of carbon, hydrogen, nitrogen, and oxygen—resembles an ugly lollipop. Ugly lollipop in shape and ugly lollipop in function, dopamine famously regulates how we respond to sugary rewards and painful punishments, what we anticipate and what we avoid. As a chemical in the brain, dopamine is fundamentally a messenger between neurons. Communication between neurons occurs at the synapse, the gap between nerve cells. To stimulate another neuron—and pass along a message, a thought, an electrochemical wave—the presynaptic neuron will deposit sacs of neurotransmitters such as serotonin or glutamate or dopamine into the synapse. Dopamine molecules are tiny but mighty. Each sac, called a vesicle, contains thousands of dopamine molecules, ready to swim across the open gap and bind to the next neuron.

Once floating in the synapse, the neurotransmitters will attach to receptors lodged in the membrane of the postsynaptic neuron. Dopamine molecules bind to special dopamine receptors that are perfectly shaped to receive them. Like a key designed to match a specific lock. Once activated by dopamine, these dopamine receptors can instigate electrochemical signaling cascades, called action potentials, that pass on the biological messages that come to constitute our consciousness and our learning, our expectations and our longings.

Dopamine fits two kinds of receptors, which have been not-so-creatively named by scientists as D_1 and D_2 receptors. Each receptor class has slightly different physiological properties, affinities, and localization within the brain. D_1 receptors are proportionally more abundant in the prefrontal cortex, the frontal area of the brain commonly linked to goal-directed decision-making, cognitive control, and high-level reasoning. In contrast, D_2 receptors are concentrated and expressed in the striatum, a snail-shaped cluster of neurons that constitutes the largest recipient of dopaminergic nerve ends in the midbrain. The striatum computes expectations about the consequences of our actions. It encodes the associations we forge between rewards and stimuli as well as between rewards and the actions we take to achieve them.

WHISPER: I hope this lesson in biology is relevant to ideology . . .

WHISPER BACK: Yes, I promise. Just wait.

It is estimated that there are between 230,000 and 430,000 dopamine neurons in the human brain. Scientists consider this to be a small number, given the presence of over 80 billion neurons encased in our skulls. In this galaxy of star-shaped neural bodies and fiery fibers, a quarter of a million dopamine neurons is almost nothing. Merely a drop in the brain bucket.

Yet the low quantity is compensated by high potency. A single midbrain dopamine neuron has axons—the long nerve fibers that extend away from the cell body toward other neurons—that can stretch for up to four meters. Imagine the four meters of elongated neuronal fibers belonging to a single neuron bunched up and coiled so tightly; some folding into spaces smaller than a pine nut or a sesame seed, and others extending far from the central nervous system to the periphery, reaching our eyes, kidneys, gut, and heart, our spongy earlobes and the tender arches in the soles of our feet.

Each dopamine neuron can affect tens of thousands of other cells, synchronizing with others to produce signals that propagate and lead to action. The dopamine system's scale in time and space is astounding: dopamine neurons pass along signals in fractions of milliseconds and these signals are transported along neuron-to-neuron connections whose total distances could easily rival the height of the tallest Amazonian tree. All of this transmission, this fast-paced, microscopically packaged electricity, is happening within each human body, *constantly*. A blooming and buzzing coordination.

Each dopamine neuron is beautifully long, generously branched, and powerful. And it is dopamine's majestic—and at times terrifying—power over our behavior that has kept scientists chasing after the dopamine rush of discovery, hoping to uncover more about how dopaminergic mechanisms work in different individuals to different ends.

For a long time, researchers have hypothesized that individual differences in cognitive flexibility may be neurochemically rooted in dopamine. Preliminary evidence for the dopamine hypothesis of cognitive flexibility originated from studies indicating that medications that modulate the level or the transmission of dopamine can treat a range of psychiatric conditions marked by rigidity, compulsivity, and unshakable beliefs. Experiments with "knockout" mice—mice for whom a particular gene has been silenced—revealed that disrupting dopamine circuitry by altering gene expression led to the formation of rigid habits and a difficulty to shift when old rules must be abandoned. Such knockout mice were more susceptible to addiction and would continue to perseverate after the intoxicating reward was eliminated; rigidity persisted even when the addicted mice had to endure electric shocks every time they pursued reward. The dopamine systems of mice, men, and women appear to shape the capacity to switch behavioral responses to external cues and changing situational demands.

As schoolchildren learn in their first human biology class, every protein produced in every cell of our body—including enzymes, neurotransmitters, and receptors—is transcribed from the code in our genes. Each gene is a sequence of nucleotides containing one of the four elemental bases—adenine, thymine, cytosine, or guanine—that together allow the cell to synthesize the building blocks that make up its internal machinery.

Some genes contain slight variations—a note of adenine instead of a thymine, or a guanine replacing a cytosine—and the presence of such variations leads to differences in traits between people. If a gene has variations in a single nucleotide, this is called a single nucleotide polymorphism (SNP), pronounced "snip." A SNP can affect whether we are born with green eyes or brown, teeth that are crooked or straight, eyelashes that are short or luscious. Even differences in our taste perception are partially explained by these "snips," which govern the degree to which nerve cells connected to our tongues express receptors that can "perceive" the molecules designating sweetness, bitterness, salty flavors, or fatty umami. These genetic differences make some individuals *supertasters*, who experience bitter flavors more intensely than most, while others are less-sensitive discriminators, or even *nontasters*. The degree to which we enjoy earthy coffee, bitter spinach, or tart citrus undertones is thus a matter, partly, of genes.

In the same way biologists map SNPs that dictate physical attributes and sensory preferences, scientists have discovered genetic variations that affect how dopamine is distributed, released, and metabolized throughout the brain.

One of the most famous genes known to neuroscientists is the catechol-O-methyltransferase gene, or COMT for short. Discovered by Nobel laureate Julius Axelrod in 1958, the COMT gene helps to regulate levels of dopamine in the prefrontal cortex. In particular, the COMT gene holds the recipe for the production of an enzyme responsible for over 60 percent of the breakdown of dopamine in the prefrontal cortex. If the COMT enzyme is abundant, it will "clear away" any leftover dopamine that floats in the synapse, thereby keeping baseline dopamine levels low. If the COMT enzyme is scarce, then dopamine will build up and linger. If residual dopamine is not swept away fast enough, this will lead to high levels of baseline dopamine, which can continue to stimulate receptors in the membranes of dopamine neurons in the prefrontal cortex.

Located on each person's chromosome 22—incidentally one of the smallest and daintiest chromosomes we have—the COMT gene comes in two forms: the *Met* allele or the *Val* allele. Since each person has two copies of each chromosome, one inherited from each parent, we can carry either two *Met* alleles, two *Val* alleles, or one of each. A person's particular combination of these two alleles, their *genotype*, makes a big difference to how dopamine is metabolized and regulated in their prefrontal cortex on a day-to-day basis.

The *Val* allele has four times the enzymatic activity of the *Met* allele, meaning that individuals with the high-activity *Val* allele experience more elimination of dopamine by COMT. The *Val* allele's higher COMT activity means less dopamine hanging out in the synapse, and so less active dopamine neurotransmission happening in the prefrontal cortex. On the other hand, individuals with the low-activity *Met* allele—for whom COMT is *not* clearing away dopamine very thoroughly—will have high levels of dopamine sticking around in the synaptic gap, ready and eager to forward electrochemical waves to more dopamine neurons.

WHISPER: Are we nearly there yet?

WHISPER BACK: Where?

WHISPER: At the end of all this detailed science! I get it. Some people have more dopamine orbiting their prefrontal synapses and others have less. So what?

The highly dopamined *Met*-carriers tend to perform better on complex cognitive tests, such as tests requiring inhibition, reasoning, and keeping many items in mind at the same time. This has led neuroscientists to wonder whether *Met*-carriers are naturally more flexible thinkers. But there is the challenge of specificity. Neuroscientists have long debated whether the COMT gene predicts a person's flexibility or captures a more generic cognitive ability that facilitates many kinds of mental processes, regardless of whether they require adaptability to change or not. For a precise explanation, we need to dissociate mental *flexibility* from mental *ability*.

Cognitive ability is a tricky variable because high intelligence can obscure interesting patterns in the data. At times, individuals with high cognitive ability can compensate for other weaker cognitive processes. I found this when I studied intellectual humility: individuals who were highly intelligent, even if they were cognitively rigid on the Alternative Uses Test, displayed a balanced receptivity to evidence and alternative perspectives. So in my study on the genetics of cognitive flexibility, I excluded participants with high intelligence in order to make sure that I was focusing on cognitive flexibility. I did not wish to confound or confuse flexibility with general cognitive ability.

In looking at the COMT genotypes of my participants, I found that cognitive flexibility on the Wisconsin Card Sorting Test was better for *Met*-carriers than for individuals with the *Val/Val* genotype. In other words, individuals who generally have more dopamine marinating their prefrontal cortex are better able to adapt to the changing rules in the game. This pattern was sustained when controlling for age, gender, and IQ. The specificity existed: flexible thinkers tend to have higher levels of prefrontal dopamine.

But prefrontal dopamine is less than half of the story. Dopamine neurons in the prefrontal cortex are far outnumbered by the ones in striatum. Experiments with animals illustrate that if dopamine is depleted in the striatum, the animals will struggle with tasks that require learning a rule and then reversing it. Adaptability may hinge on the dopamine in our midbrain too.

I explored individual differences in baseline levels of striatal dopamine by studying the gene that codes for the D_2 dopamine receptor that permeates the striatum. A Finnish research group had convincingly shown that a SNP in the DRD2 gene can differentiate the amount of available D_2 receptors in the striatum. At a particular location in the base sequence of the DRD2 gene, some people will have a cytosine (C) and others will have a thymine (T)—and since each person has two copies of each, some people will have a C/C genotype, others a T/T genotype, and some will have a C/T genotype. Individuals with a C/C genotype have fewer D_2 receptors available in their striatum than their peers with a C/T or T/T genotype. Fewer available D_2 receptors result in more dopamine molecules becoming concentrated in the synapse, waiting for their turn to bind to the scarce receptors. So individuals with a copy of the C allele are thought to have higher levels of baseline striatal dopamine than individuals with copies of the T allele, who should have lower concentrations of striatal dopamine. I discovered that C-carriers, with high striatal dopamine, performed poorly on the Wisconsin Card Sorting Test, whereas T-carriers, with low striatal dopamine, were adaptable and reactive to change.

And when I combined the genotypes coding for *prefrontal* dopamine and those coding for *striatal* dopamine, I found that the most rigid individuals are those who have *both* low prefrontal dopamine (*Val* alleles for the COMT gene) *and* high striatal dopamine (C alleles for the D_2 receptor gene) according to their genotypes. This genetic profile puts people at risk for mental rigidity.

This study was able to resolve debates about how dopamine genes interact with each other to produce rigid traits. The large participant pool allowed me to investigate nuanced interactions between genes that are usually impossible to discern with smaller samples. Although this was a single study—and so requires further elaboration and replication—it was consistent with meta-analyses that combined the results of multiple scientific endeavors.

The way dopamine populates our reward-learning circuitry can foreshadow how adaptably we behave. And so behind personal rigidities, genetic differences may be lurking too.

WHISPER AGAIN: So it's all fixed? Predetermined!

WHISPER SOFTLY BACK: What is?

WHISPER [*indignant, scandalized*]: If my dopamine is concentrated in the midbrain striatum and not the mighty prefrontal cortex, then I am doomed to rigidity?

WHISPER LESS SOFTLY: No, genetic influences do not mean that outcomes are fully determined. Instead, genes shape the potential expression of—[*interrupted*]

WHISPER [*A tone of outrage tries to disguise a bubbling fear*]: And if you are lucky and your baseline dopamine levels in the prefrontal are high, and low in the striatum, then you are guaranteed flexibility for life?

WHISPER BACK [*irritated*]: No, that's not what I said—

WHISPER [*more frenzied now*]: So we have no free will, no agency, our hands chained by the double helix's twisted and cruel fate . . .

WHISPER BACK: You're spiraling!

WHISPER [*with theatrical agony*]: . . . Our dogmatism is determined by our dopamine in a savage destiny we cannot control or escape or—

A CRY: Stop!

[*The melodramatic voice turns to look, surprised and somewhat relieved by the interruption.*]

[*Regaining composure, the scientist's voice speaks again.*]

WHISPER BACK: Genes interact to create *possibilities*, ranges of probabilities, skewing biological mechanisms toward different patterns of functioning. But there is a difference between *potentiality* and *actuality*.

[*A pause.*]

WHISPER: So you are saying that the path between our genes and life outcomes is not predetermined?

WHISPER BACK: Exactly, there are many possible routes and trajectories.

WHISPER: But if that's the case, what factors affect whether our genetic dispositions will be expressed or not?

WHISPER BACK: Well, obviously that depends on circumstance . . . and chance . . .

WHISPER [*horrified*]: Chance?

WHISPER BACK: . . . And choice.

WHISPER [*eyebrows raised*]: Choice?

Rigidity does not transpire out of thin air, nor does it originate from the wellspring of a metaphysical soul. Individual differences in inflexibility emerge from biological markers that interact with each other. No single dogmatic gene exists. All psychological characteristics are shaped by a plurality of genetic mechanisms. In the same way that there is no single gene determining charisma or sense of humor or sadistic tendencies or how easily we cry or giggle, our belligerence cannot be explained by a unitary gene.

Genes interact with each other in tangled forms. Like the agonizingly complex relationships between people—friends, lovers, and strangers—interactions between genes occur in varied assemblages. Some genetic effects amplify one another, leading to pronounced traits, whereas some genetic effects cancel each other out. Some genetic effects multiply and accelerate each other, leading to nonlinear accumulation of effects or to cases when interactions between genes produce phenotypic traits that are not visible in parents (imagine the gifted genius born of apparently mediocre parents).

At times a gene's downstream effects can even compensate for variation in another gene. One such compensatory interaction emerged in the genetic investigation of flexibility: having the low-prefrontal-dopamine genotype *and* the high-striatal-dopamine genotype led to cognitive rigidity—both were necessary!—but having only one was not necessarily sufficient. If a person is endowed with one flexibility-prone genotype and one rigidity-prone genotype, the flexibility-prone genotype will frequently dominate. This compensation effect hints at the inherent plasticity of our brains: when one system struggles to facilitate adaptability, another system substitutes.

Human biology consists of dynamic processes that cooperate and collide with each other, rendering each person a unique constellation of interactive, changing biological forces that react to the outside world. Indeed, it is this very reaction to the environment, the *interaction* with the environment, that eventually constitutes who we become. Each person has the scope to become many versions of themselves: their genotype can predispose them to certain cognitive styles or abilities, but ultimately their *phenotype*—their observable personality, physique, outcomes, successes and failures, habits and fears—emerges from the interaction of genetic predispositions and salient experiences.

When scientists uncover genetic effects, they are also on the lookout for epigenetic effects. Epigenetic effects refer to the fact that the expression of genes is not completely fixed over time: whether a gene is activated, subdued, or suppressed depends on a person's lived experience in the world. Epigenetic modifications are not alterations of the DNA itself. Epigenetic modifications are alterations of the factors that promote or inhibit the transcription of DNA into molecules. The prefix *epi-* denotes that these alterations sit "over," "on top of," "above" the DNA—and so epigenetic changes caused by experience can occur throughout our lives, giving a collage-like texture to our sensitivities. Experiences get inscribed into the body for better or for worse, and often for better *and* for worse, depending on the situation. An acute sensitivity can breed discernment and creativity in one context, and in another it can lead to vulnerability and distress.

I believe that when we dissect ideological thinking—the style of thinking characterized by rigid adherence to a dogma and a rigid social identity—we need to consider the emergence of ideological rigidity in epigenetic terms. That is, we need to develop an "epigenetics of extremism."

An epigenetics of extremism tries to project how certain biological and cognitive vulnerabilities to rigid thinking fare in different environments. It allows us to ask questions about the interaction between baseline psychological traits and ideological experiences. How would a person with a flexibility-prone genotype fare in a dogmatic and absolutist environment with repetitive rituals and strict rules for behavior and imagination? What would happen to the rigidity-prone genotype in such a dogmatic setting? And what about the rigidity-prone genotype in a secular, liberal, open, tolerant, anti-racist, anti-sexist environment—would they be less rigid than their imaginary doppelgänger in the dogmatic, prejudiced, authoritarian environment?

An epigenetics of extremism encourages us to challenge the premise of the chicken-and-egg problem. It moves away from the notion that dogmatism is either the product of our biological vulnerability *or* the effect of indoctrination. Instead of considering mutually exclusive causal possibilities, we can look at dynamics and interactions—how different brains are seduced and affected by different orthodoxies. We can ask who is most susceptible to the effects of rigid ideological doctrines and how their brains and bodies may be changed as a result. We can turn our attention to the question of consequences.

WHISPER: But, wait. [*suddenly timid*] Why are we whispering?

WHISPER BACK: [*an uncharacteristic pause*] Because there is an uncomfortable idea we are insinuating . . . an unnerving implication hiding just below the surface . . .

Part IV

CONSEQUENCES

13

DARWIN'S SECRET

When I think of his restlessness, I imagine it at nighttime. The stillness of the dark contrasted with the pulsations of his darting thoughts. A cacophony of chattering feelings all wrapped up in the softness of shadows. Silent yet thundering heartbeats marking the insomniac's passage of exhausted time.

But it could have also happened the other way. The juxtaposition in reverse. A quiet anguish, calm and considered, felt in the midst of the household's morning noise and unmelodic movements, cutlery clattering senselessly, surrounding this old man as he resigns himself, sorrowfully, ambivalently, to the intractable conflict of his life.

If it was midday, I imagine Darwin looking through the study's heavy glass window and seeing his wife, Emma, in the garden, reading from the New Testament. Hers was a well-worn and thickly annotated copy, layers of notes deposited on top of each other like coarse geological sediments archiving a history, harboring relics and remnant secrets to be excavated later. Emma would be fingering the Gospel of John, a section she had described to Charles as "so full of love . . . & devotion & every beautiful feeling."

Nearly forty years earlier, in the jittery days of their engagement,

Emma had worried about Charles's religious skepticism, lamenting that it would create "a painful void between us." She urged him to read the Gospel of John's thirteenth chapter, praying that its message of charity and virtue would cement Charles's faith and steer him away from the dangers of atheism. "May not the habit in scientific pursuits of believing nothing till it is proved," Emma had written to him, "influence your mind too much in other things which cannot be proved in the same way."

Religion was to be a precondition of their union. Doubt could be its undoing.

But now it is a cool July night in 1876, in Kent, in a country house that Charles Darwin has shared with Emma for nearly four decades. He is no longer a young man flushed with adoration and the energy to persuade. Charles is a reflective white-bearded theoretician of sixty-seven, revisiting old letters and correspondences, annotating them with his own kind of reverence, as he crafts a definitive autobiography that would testify to his voice and his truths.

After decades of discovery and scientific debates, Charles Darwin is turning inward, returning to past promises. Promises he realizes now he did not and could not keep.

Back in his youth, he had initially consented to his father's urging to preserve marital harmony with the illusion of faith. But, impulsively, the thirty-year-old Charles confessed. Confessed his skepticism to Emma's horrified ears. He had been too optimistic, too convinced by love to realize that Emma would not relent. Charles—that intellectual giant!—was forced to capitulate, to swear he would extinguish his doubt about divine design or godly omniscience. An implicit contract was struck and signed. An agreement he would arduously try, and fail, to honor.

Now, toward the end of his life, all the promises and suspicions that Charles expressed to Emma at the beginning of their courtship—all the assurances and misgivings—are tormenting him again.

I imagine him lying awake at the cusp of a yellow dawn, head facing starward and hands clasped around his neck in his customary position, feeling at once tired and animated, knowing that he cannot contain it any longer. He *must* write the ideas that have been racing through his head, unadorned by euphemisms or delicate evasions.

I see him turning over in his bed and gazing at his beloved Emma sleeping soundly. *What would Emma think of his troubled ruminations, if she knew his final conclusions after a lifetime by her side?* I picture Darwin carefully creeping out of bed and heading toward his study, excitement and anxiety swirling in his fingertips. With a dark, sharp-tipped pencil he writes:

> [We must not] overlook the probability of the constant inculcation in a belief in God on the minds of children producing so strong & perhaps an inherited effect on their brains not yet fully developed, that it would be as difficult for them to throw off their belief in God, as for a monkey to throw off its instinctive fear & hatred of a snake.

In fewer than a hundred words, Darwin articulated a whirlwind of a hypothesis. A hypothesis of multiple parts, tiers, and analogies. Yet this is a hypothesis that would be promptly erased from Darwin's autobiography upon his death in 1882. When Charles's son Francis prepared the autobiography for publication in 1885, he received a letter from Emma:

> *My dear Frank,*
>
> *There is one sentence in the Autobiography which I very much wish to omit, no doubt partly because your father's opinion that* all *morality has grown up by evolution is painful to me; but also because where this sentence comes in, it gives one a sort of shock—and would give an opening to say, however unjustly, that he*

considered all spiritual beliefs no higher than hereditary aversions or likings, such as the fear of monkeys toward snakes.

. . . I should wish if possible to avoid giving pain to your father's religious friends who are deeply attached to him, and I picture to myself the way that sentence would strike them . . .

Yours, dear Frank,

E. D.

It was only in 1958, over eighty years after Darwin committed these ideas to paper, that his granddaughter Nora Barlow helped publish a revised edition of the autobiography, which filled in the omissions of previous versions. However, by that point, psychologists and political scientists were not interested in searching for inspiration in the historical footnotes of an obscure memoir. They had learned to quote from Darwin's theories about evolution and natural selection or from his treatises on the variety and substance of human emotions. But as far as psychologists knew back then—or are aware of now—Darwin had no clearly stated politics, let alone thoughts about human ideology. Hidden for so long, his provocative idea about indoctrination was blurred and buried.

Darwin's hypothesis, deleted from scientific history through Emma's editorial intervention, considers what would happen to a child's brain if it experienced "constant inculcation" by religious orthodoxy. To inculcate is to force upon, to stamp in, to tread on, to trample, to impress, to instill, to insist, to crush with the heel (*calx* in Latin, the calcaneus bone, the origin of the Latin *inculcare*). When Darwin writes of "constant inculcation," he is thinking of religion as enforced from outside rather than discovered within. He is imagining that religion could be a system, not merely a source of love and devotion and spiritual feeling, but a system that is impressed upon people, *children*, by force. A system learned by supple, impressionable young brains. A system learned through the same mechanisms by which animals learn to fear and to hate, to attach or to escape.

Darwin reasoned that the consequences of religious inculcation and of repetitive religious practice could be strong, long-lasting, and as biologically real as animal instincts and genetic inheritance.

It was one of Darwin's most overtly political, damning statements. No wonder the words were forcibly removed from the pages of his autobiography. There were too many people such a sentiment would offend, a legacy too precious to jeopardize. In letters to his son George, Darwin mentioned his own self-censorship throughout his career, urging him to "pause, pause, pause" before publicly sharing his thoughts on the fallacies of religion. "Direct attacks on Christianity," Charles advised, "produce little permanent effect: real good seems only to follow from slow & silent side attacks."

What a cunning strategist Darwin was! If only he knew that when he finally decided to lift the lid on his secret hypothesis, it would be swiftly shut by Emma's equally cunning hand.

Darwin's hypothesis weighs how difficult it would be for children "to throw off their belief in God": how possible it is to learn religious beliefs and then to undo the learning, to wiggle from the cast of orthodoxy and run.

But perhaps Darwin's vision of religion was too constricted or reductive. Maybe religious affiliation is in fact linked to a kind of openness to the world—an understanding that there is more to the cosmos than meets the eye. Rather than a cast, religion can be viewed as a scaffold to reach higher meanings and forms of happiness. For the believer, life is given a structure and rituals that punctuate a life's successes and sorrows and define an alphabet of ways to seek solace, treat heartbreak, and experience the sublime.

To address Darwin's question about the effects of religion on the brain, it is necessary to engage with empirical science rather than purely theoretical speculation.

I wonder whether, as a famous experimentalist, Darwin contemplated which potential tests could be devised to verify or falsify his hypothesis. Granted, children were not Galapagoan finches whose jet-black beaks could be closely appraised. Nor were children akin to tropical orchids whose petals could be stroked and scrutinized as part of the botanical studies he conducted in his English garden's "outdoor laboratory." But Darwin did not recoil from complexity or the evasiveness of time. Thinking about the long time periods during which a creature could change, evolve, or respond to its environment was Darwin's trademark move.

How would a scientist assess the degree to which religious dogmas shape the brain? Can we tease apart the components of "constant inculcation" on the minds of children—the experience of repetitive religious practice and the unique sensitivity of the developing child? Which is more important for the formation of religious convictions: the regularity of engagement with religious codes, its timing in the lifespan, or the intensity and fervor of the religious practice?

The ideal method is a longitudinal study. A recipe of verbs for the scientist to follow. TRACK many babies from the cradle to the playground and on to the seasons of adulthood. MONITOR them *closely*, and the interplay between causes and effects, dispositions and environments, nature and nurture, will manifest itself. TRACE how circumstances, chances, and choices come to constitute the person from their earliest beginnings to their last ruminations. CAPTURE the person from all angles: past and future, conscious and unconscious, private and public. TREAT the person as an object that you can subject to meticulous measurements.

A lifetime of data points, of patterns, of *befores* and *afters*. It is the optimal method.

But a scientist rarely has a lifetime to await the results.

So if a scientist can't or isn't patient enough to wait a lifetime, then she can look for differences between lifetimes. When we can't fast-forward time, we can instead look backward at the memories evoked by different lives—differences in upbringing and outlooks that can expose the extent to which ideological environments shape the developing brain.

Religion is one of the strongest and most memorable forms of an ideological upbringing. This is because religion is highly performative, both as a doctrine and as an identity. Religiosity is not merely an internal conversation about moral codes, cosmic destinies, and metaphysical claims. The interiors of religion are clearly and purposefully externalized. Unlike the plausible secrecy of prejudice or the semblance of confidentiality we experience with regard to our political choices (voting behind a flimsy curtain separating one booth from another—akin to a hospital's shocking definition of privacy), religion is worn on the sleeve. Symbolic necklaces adorn praying chests. Bindis or pottus dot dutiful foreheads. Cloths cover pious heads. Unshorn hair is knotted beneath a crowning turban or curled into bouncy ringlets as a sign of religious identity. Some colors are consciously shunned and others deliberately shown. Bare skin is at times necessary for sacred rituals, ready for temporary decoration or permanent alteration. At other times the skin is considered immodest or impure.

Religious doctrines encourage the believer to embody their faith, and so the experience of religious participation is not quickly forgotten. Even years after the fact, it is possible to recollect the strength of influence religion had over an upbringing, its effects on the skin. As a result, there is retrospective clarity. Ask an adult how frequently their family attended religious services, performed acts of prayer, and expressed devotion to a religious ideology, and most adults can categorize their upbringing with confidence. Religion is tailored to incorporate children from the earliest stages. Religious rituals revolve around birth, circumcision, baptisms, communions, and confirmations, clear rites of passage through the milestones of childhood, adolescence, and adulthood—gateways in and out of the mortal world, a time that is bookended by the mist of the supernatural. Children are explicitly taught religious imperatives as well as rituals that they can practice privately and share publicly with a community of fellow believers. It is difficult to forget the emotions of collective religious experiences—revelations washed with awe, glee, and solemnity; pilgrimages borne together; the tingling trance or dread or harmony stirred by worship.

By probing individuals' memories of the religiosity of their upbringing, a history can be created, a retrospective record that allows the scientist to time-travel without waiting a lifetime. By studying the differences or similarities between a person's current religiosity and their past, we can observe continuity and conversions. This chronology is important because people's religiosity is not fixed—it can change over time.

I was curious to explore how individuals' religious affiliation, religious upbringing, and levels of practice, prayer, and engagement were related to their cognitive flexibility. In a study of over 700 people, I found that cognitive flexibility was linked to religious disbelief. The effects were large and consistent across different tests of both reactive and generative flexibility—that is, in tasks that entailed adapting to changing situations and in tasks that called for spontaneous invention. As with other ideologies, strong religious conviction was related to greater cognitive rigidity.

Within religious belief there are gradations of commitment. And so a question blushes into view: whether all religious adherents have the same cognitive profile. The data suggests that the answer is no: there is substantial variation in the mental flexibility of religious believers. Some of this variation is explained by people's levels of conviction and active commitment. The more frequently people practice their religion, engage in repetitive rituals, pray, and attend religious services, the more rigid their scores tended to be on the neuropsychological tasks. And the less frequently and ardently people practiced their religion, the higher their flexibility scores.

One of the most striking results was that nonbelievers—regardless of whether they grew up atheistic, agnostic, or devout—exhibited the most adaptable behavior on the Wisconsin Card Sorting Test and the most flexible solution sets in the Alternative Uses Test.

It is what a person *chooses* to believe—rather than how or where they were raised—that is most emblematic of their cognitive style.

Does this mean that a person's upbringing and history are irrelevant? Is it only the present that counts? On the contrary, a close analysis of the data reveals that upbringing is important. We can glean clues from the *converts*, the two groups of people who changed their religious status: the people brought up religious who left the religious ideology and the people brought up secular who entered a religious ideology by choice.

To me, these are the most interesting people: the ones who defied their parents' expectations and drove their lives in a different direction. In the data, I discovered that those who *left* the religious ideology achieved the highest flexibility scores. In some flexibility tests, the newly secular individuals even had an edge over the people who had always been secular. And the people who converted *into* a religious ideology attained the lowest flexibility scores on average, at times more rigid than people who maintained a religious attachment all their lives.

There are still many open questions for future research. The issue of causality—the chicken-and-egg problem—is not settled. It is difficult to tell whether people who are cognitively flexible tend to gravitate toward secular worldviews or if choosing to partake in secular environments leads to more flexible mindsets. Moreover, human civilizations have assembled a kaleidoscopic set of religious scriptures and practices, and so different faiths may invite and create different cognitive traits. Recording people's entry into and exit from a diverse range of religious sects is therefore paramount to understanding the psychological basis of religious convictions and conversions.

The writer and psychoanalyst Adam Phillips writes astutely of such changes: "Conversion tends to be a cure for skepticism; a narrowing of the mind that frees the mind. Frees it, one might say, from [a] kind of complexity—the relishing of complication, of diversity, of contradiction." Ideology demands that reality align with ideals, that ideals align with each other, that everything be unified and organized.

So it makes sense that converts *out of* an ideology are highly flexible—they must dislodge themselves from an entire universe of principles that have organized their lives. To detach from an ideology requires a painful ripping away from communities that previously offered intellectual and emotional support. Social rules are exchanged for self-rule. As a result, revolt against ideology requires extraordinary flexibility.

Many religious believers would disagree with this characterization. For them, religiosity evokes an openness toward a more expansive world. A world colored by divine beauty and purpose. A world of "depth," according to the philosopher Paul Tillich. A world of "surrendering," for Simone Weil. A majestic "leap" for the thinker Søren Kierkegaard. Religious revelation in this sense speaks toward *an open heart*, an openness to a spiritual world with invisible dimensions and meanings.

But I believe the language of "openness" misleads. It muddies the difference between living in a way that is amorphous and living in a way that is rigidly defined. By using "openness" to describe conversions into an ideology, we are blurring the lines between education—which opens the world—and indoctrination—which closes it.

In religion, we find the ultimate fulfillment of the brain's predictive and communicative patterns. Someone is always present to listen, to story-tell, to explain. Mind and meaning are attributed to every inexplicable event. Religion postulates intelligent powers, supernatural spirits, and mystical events in a way that uniquely speaks to our brain's tendency to see agency in random events and to imagine that chance encounters are underpinned by intentions. This is a very human reaction. Even the most secular mind can become superstitious under stress, suddenly gravitating to lucky omens or believing in the possibility of manifesting dreams or jinxing fears into existence.

Yet by promoting orthodoxies rooted in hidden entities, religion creates a conflict between sensation and reality. The philosopher David Hume wrote of this tension in his 1757 book *The Natural History of Religion*, observing that "however strong men's propensity to believe invisible, intelligent power in nature, their propensity is equally strong to rest their attention on sensible, visible objects; and in order to reconcile these opposite inclinations, they are led to unite the invisible power with some visible object." To reconcile expectations of supernatural events with the absent evidence for them, our brains' rational processes update our beliefs to infuse real sensory experiences with an extrasensory meaning.

Even Darwin was not blind to the beauty of reverence or its optical power. In his autobiography, in parts left uncensored—the parts Emma could not convince the editors to touch—he contemplated:

> In my Journal I wrote that whilst standing in the midst of the grandeur of a Brazilian forest, "it is not possible to give an adequate idea of the higher feelings of wonder, admiration, and devotion which fill and elevate the mind." I well remember my conviction that there is more in man than the mere breath of his body. But now the grandest scenes would not cause any such convictions and feelings to rise in

> my mind. It may be truly said that I am like a man who has become color-blind, and the universal belief by men of the existence of redness makes my present loss of perception of not the least value as evidence. This argument would be a valid one if all men of all races had the same inward conviction of the existence of one God; but we know that this is very far from being the case. Therefore I cannot see that such inward convictions and feelings are of any weight as evidence of what really exists. The state of mind which grand scenes formerly excited in me, and which was intimately connected with a belief in God, did not essentially differ from that which is often called the sense of sublimity; and however difficult it may be to explain the genesis of this sense, it can hardly be advanced as an argument for the existence of God, any more than the powerful though vague and similar feelings excited by music.

Darwin's ruminations are imbued with colors, sounds, sensations of the sublime. Religion is a perceptual framing—a frame of mind that alters the intensity of what we sense as well as what we do not notice at all. Supernatural meanings are layered on top of sensory experiences. Discrepancies and doubts are muted into silence. In losing his faith, he was "like a man who has become color-blind"—no longer seeing a supernatural dimension to life that he previously wished was there—and, at the same time, a man who could now see the world anew: more flexibly, more honestly, more freely.

14

POLIPTICAL ILLUSIONS

Once seen, the provocative image could not be unseen. The German cartoonist who drew it had no idea of the wrath, of the roaring confusion, such a picture would produce. The controversy nearly caused riots in the streets. Philosophers were baffled and began filling their texts with accusations—claiming this cartoon had upended everything, turned reality upside down. The arresting picture had cast shadows over the entire basis of human perception, even of human morality, and so everything would need to be reassessed. The gray sketch revealed how easy it is to trick the mind, to polarize the populace, to see reality not as it is but rather through impressions that are hazy and fragile—and wrong.

The infamous cartoon was the duck-rabbit, or rabbit-duck, illusion. It was initially published in 1892 almost as an afterthought, covering only an eighth of a page in the back of a magazine, where sudokus and other riddles now sit. The reaction was raucous. The term "polarization" would only be transferred from physics to politics in the twentieth century, fifty years after the cartoon's publication—but here, in spectators' searching eyes, was polarization in all its messy glory.

Within a month, the renowned American political magazine

Harper's Weekly reprinted the niche Bavarian cartoon, disseminating it across the United States. It was the nineteenth-century version of going viral.

Etched in thin black strokes, the ambiguous drawing resembles a pouting duck at one glance, or a long-eared rabbit at another. In shifting from one impression to the next, the duck's split beak transforms into two rabbit ears. The duck's face pointing to the left suddenly appears to be a rabbit staring blankly to the right, its bunny ears relaxed backward.

Some see the rabbit first and the duck second, while others detect the duck and struggle to recognize the rabbit. Often, we need a trigger, a push, a conveniently placed caption or a mumbling neighbor in order to be told that there is another animal lurking in the figure. We might squint, tilt our heads, move the cartoon closer and then further away before we can break free from our first impression and see reality afresh. Otherwise our attention gets locked by the contours of our first impression, and we never notice the picture's duality, its double-sidedness.

Some viewers will flip between duck and rabbit more easily than others: people with heightened cognitive flexibility on the Alternative Uses Test have been shown to switch between the faces of the duck-rabbit illusion with greater ease. Context and memory cues matter too. A charming study of over 500 Swiss participants conducted at the entrance gate of the local zoo found that in October people are most likely to report seeing a duck, whereas around Easter time in spring viewers will disproportionately report seeing a rabbit. The Easter effect is strongest in children under ten, for whom Easter Bunnies are, understandably, most salient. Adults' and adolescents' first impressions are less shaped by the Easter Bunny's presence or absence; *it is a less magical time.*

When we make the switch from seeing a duck to seeing a rabbit, we experience the warmth of the eureka moment—the "aha" of insight and revelation—and a giggle of surprise. We laugh because we have learned something new. We laugh because we have been challenged in a small but significant way. Our assumption that there is a single representation underlying any given image, that the world is stable and irreversible, has been violated. We have been pranked. And so we tend to laugh, to share the joke with our friends, to find out what version of reality they intuit first.

"What do YOU see?" we ask, with a smile of compassionate superiority.

Since the image mocked us and disclosed our faults, we now use it to tease and torment others. After all, "a thing is funny," George Orwell observed, "when—in some way that is not actually offensive or frightening—it upsets the established order. Every joke is a tiny revolution."

For one of the world's most famous philosophers, however, the duck-rabbit illusion was not a tiny revolution—it was a monumental one.

In his celebrated 1953 treatise *Philosophical Investigations*, the acclaimed Austrian philosopher Ludwig Wittgenstein wrote at length about the duck-rabbit illusion, fascinated by its tantalizing ambiguity. He even drew a rudimentary version of the cartoon into his book's fragmentary arguments. An odd jolt of lightness in a sea of philosophical seriousness. Wittgenstein argued that the duck-rabbit illusion demonstrated that when we see an object, we are not necessarily seeing it for what it is. Visual experience is imbued with judgments and biases, attention randomly captured and then held hostage.

Philosophically, the revelation was a travesty. If ambiguity misleads us so easily, camouflaging truths and leading us astray, our access to the world is faulty and unreliable. We comprehend only a fraction of reality and neglect its entirety. *Imagine how many things we might regularly miss and misinterpret.*

The phenomenon is even more troubling when we realize that

prior knowledge and familiarity do not protect us against selective perception. Even once we have discovered the illusion's dual images, it is impossible to see both interpretations simultaneously. The image is both duck and rabbit, but never precisely at the same time. There is a pictorial rivalry. Look at the doodle for long enough and it is clear that even when we know that both animals are buried in the ink, at no moment can we see both animals at the same time. It is a bistable illusion, reversible and fickle.

In Wittgenstein's typical aphoristic and conversational style, he asks, "what is different: my impression? My point of view? . . . I describe the alteration like a perception; quite as if the object had altered before my eyes." The shift feels radical, as though the world itself has changed, not our minds. We experience a similar dawn in our everyday lives when, roaming in a crowd, we unexpectedly recognize a familiar face. "I contemplate a face, and then suddenly notice its likeness to another," Wittgenstein writes. "I *see* that it has not changed; and yet I see it differently."

The switch from one image to the other was labeled by the German pioneers of experimental psychology as a *Gestalt* switch. *Gestalt* roughly translates to "whole": we shift from seeing one whole entity (a rabbit) to another whole entity (a duck). The mind does not perceive in pieces and parts—it integrates and senses the whole. Perception is not broken up into atomic particles—it is a continuous experience. This is why we see animals and faces in aimless clouds; our minds extract shape, structure, and meaning. It is also why we get stuck on a worldview and struggle to reimagine the information in another configuration. Things appear to us as unified.

For the brain, the whole is holy.

Rarely do we experience blurs or ambiguities in our vision that are left unexplained. We construct and selectively overlook inconsistencies or inconveniences. For a philosopher like Wittgenstein, this is a traumatic realization. It implies that reality may be just a mirage, a useful hallucination that our minds generate from partial truths and selected inputs. Our sense of reality is an interpretation.

Yet if there is no ground truth, how can we ever find common ground? If half of us see rabbits and half of us see ducks, there is little hope of agreeing on what reality is. This is not a mere philosophical conundrum—it is fundamentally political. We might go to war over our disagreements, sharpening our knives over conflicts in interpretation. In a millennial reincarnation of the grayscale sketch, the *New Yorker* cartoonist Paul Noth borrowed the iconic illusion and plastered the duck-rabbit image on miniature flags held by two warring tribes, all holding on to identical flags across a valleyed ridge. Lifting a horned, arrow-shaped sword, the army leader of one tribe proclaims: "There can be no peace until they renounce their Rabbit God and accept our Duck God." To the bitter end, we fight!

Readers nod and laugh. Smile knowingly. *The absurdity of it! The truth of it!*

Our interpretations of the same object, the same event, divide and polarize us. We moralize our interpretations, waging wars over simple illusions, mere ambiguity.

Visual perception is not merely a convenient metaphor for ideological perception—vision can be a scientific tool for probing the processes that lead a brain to be tempted or repelled by ideological narratives. Differences in sensory perception reflect how different brains process complex visual images or audiovisual scenes, such as bunnies that morph into ducks, red triangles flashing on the periphery of our visual field, auditory buzzes that signal when to press the keyboard and when to inhibit the response. Like optometrists, psychologists can probe the mechanisms of vision and of politics at the same time. *Which interpretation do you prefer? Which image is sharper?* A or—*click*—B? B. Okay. Now B or—*click*—C?

This research reveals that sensory processes that take place in under a second can be tied to ideological decisions that take years to crystallize and solidify. Optical illusions can become political illusions. Blending optics and politics, we can observe what I call "poliptical" convergences—parallels between our sensory preferences and political susceptibilities. How do political psychologists discover these poliptical phenomena?

There are two modes to scientific research. The first mode is the *theory-driven* approach, in which the scientist thinks deeply about the hypothesis and then crafts experiments that will falsify or corroborate the conjecture. In theory-driven research, attentive theories guide the design of tests and analyses, so that the hypothesis can be supported, rejected, or refined. Few scientific theories are ever *proven.* It is a common myth that science seeks to establish hard facts; in reality science is about the constant development of knowledge—tentative, self-questioning, curious, playful, critical, and periodically changing. Theory-driven research proceeds by asking a question, predicting the outcome, and testing whether there is good evidence for the hypothesis.

But there is also a second mode, which flips the theory-driven approach on its head. Sometimes, a scientist is in an exploratory phase of research. The scientific questions are open-ended and the hypotheses are still evolving. It is then possible to conduct *data-driven* research—research in which scientists collect rich datasets and conduct exploratory analyses to identify patterns in the data. In the data-driven mode, the scientist is not imposing their preconceptions about what the results are likely to show. Instead, the scientist lets the data speak for itself. This approach can yield unexpected results or patterns that the scientist might not even have imagined. It is also enormously useful for tackling situations in which there is a fear that researchers' biases might limit their scientific imagination by influencing the kinds of questions they ask or the favored hypothesis they concentrate on testing. When we collect large datasets, we can find answers to questions we did not know we should ask.

After I followed a theory-driven approach to map out the relationships between cognitive rigidity and ideological rigidity in the domains of nationalism, political partisanship, dogmatism, extremist attitudes, and religiosity, I wanted to know what *else* characterizes the ideological mind. What other psychological traits might emerge in individuals who believe in an ideology zealously?

Conducting an expansive data-driven study allowed me to delve into the unknown unknowns. And so I collected a dataset of over 300 American participants who had previously completed a battery of thirty-seven cognitive tests and twenty-two personality surveys. The psychological tests had taken each participant over ten hours to complete—a process they could execute at their own pace over the span of two weeks, from the serenity of their homes. Two years later, I invited these same participants to complete a series of questionnaires about their ideological worldviews. I wanted to test ideology from every direction: social and economic conservatism, nationalism, patriotism, partisan identities, support for ideological violence, participation in religious practices, resistance to evidence-based belief-updating, and attitudes toward diverse policies from abortion to welfare to climate change. This would be the most comprehensive picture produced of the psychological traits of different ideologues to date. A data-driven approach would perfectly suit such an extensive and multidimensional dataset.

I was interested to explore what we could discover when we analyzed the psychological correlates of different ideologies. Would we find psychological differences between conservatives and liberals? Would we identify parallels between dogmatic thinkers of all stripes? And which perceptual and personality traits would be implicated when we did not impose our assumptions on the data but rather followed the patterns that arose from the analysis?

A psychologist named J. Richard Simon discovered in the 1960s that a curious phenomenon occurs when people are asked to make perceptual decisions. If you teach a person that whenever they see a red circle they should press a button on their right, the location of the red circle on the screen matters. When a person encounters a red circle on the right side of the screen, the speed with which they press the right button is faster than if the circle appears on the left side of the screen. Similarly, a stimulus inviting a left-button press will lead to faster and more accurate responses if the stimulus appears on the left side of the screen. When the location of the stimulus matches the required hand action, our brains are faster to execute the response than when there is incongruence between the side of the visual field where the stimulus appears and the side of our body that must respond.

This is true for everyone. The Simon effect occurs because our attentional mechanisms tend to be skewed in the direction of our response. If you detect a ball coming from the left, it would be odd for your body to expect to try to catch it by moving to the right.

Although the Simon effect is universal, there are differences among individuals in its size. Some brains can overcome incongruity quickly and with ease. Others take longer and make more errors when facing the incongruent trials. Exhibiting a small Simon effect and being able to ignore the incongruity without trouble is not necessarily a virtue—sometimes being distracted by irrelevant features can inspire our creativity—and so these individual differences are not regarded as being either good or bad. They are simply indices of difference.

In my case, when I complete "congruent" trials where the left stimulus is in fact on the left side of the screen, and the right stimulus is on the right side of the screen, my average reaction time is 503 milliseconds, approximately half a second. When I encounter the "incongruent" trials, where the left stimulus is distractingly on the right, my average response time is 650 milliseconds—a little longer. My individual Simon interference effect is the difference between the two conditions: how much slower I am when the stimuli are on the "wrong" side of the screen, requiring me to press the left button even though the stimulus is on the right. For me, the difference between the incongruent and congruent trials was 147 milliseconds.

This time span may seem small and insignificant. Why should we care about differences of a seventh of a second? That's faster than an eye blink. Seemingly trivial. Although these are incredibly fast decisions, from the perspective of the brain these are decisions nonetheless.

Zooming into these split-second decisions can be a powerful technique. By giving participants these kinds of perceptual tasks, I was able to quantify fundamental features about how their brains attend to the world, resolve conflict, inhibit responses, and learn patterns. By focusing on such microscopic time scales, I was tapping into processes that were too fast and too unconscious for an individual to control.

When cognitive scientists analyze the results of these perceptual decision-making tasks, there are two kinds of data that are interesting: reaction times and accuracy. This can allow us to see who performs the task quickly and who performs the task well. But we can obtain even more meaningful psychological variables from such data using computational modeling. Computational modeling is an approach that takes the participant's behavior on each second-long trial and estimates what psychological processes are taking place between the moment the participant is presented with the stimulus (time zero) and the moment they press the button (in my case, 650 milliseconds).

What goes into these 650 milliseconds? Through computational techniques, we can extract each individual's *evidence accumulation rate* on a given task, which captures how quickly and efficiently a person is able to integrate the evidence presented by the task into a correct decision. Some people accumulate sensory evidence quickly and skilfully, whereas others struggle to interweave the sensory evidence into a decision. Cognitive scientists believe that this process of evidence accumulation is noisy and nonlinear, meaning that the brain doesn't immediately shoot for the right decision—it flips and flops several times before settling on an action. It is a stochastic process. Millisecond by millisecond, there are moments of growing certainty followed by hesitation or random lapses in attention. If we imagine this process like a climb up a mountain toward the peak—to the correct decision—the slope would be uneven and bumpy. Some rocks might tumble down and we may slide downward, erroneously descending to the bottom instead of hiking to the top. Millisecond by millisecond, the visual cortex communicates with the parietal and temporal cortices—where attention binds strands of information together—and the motor cortices that govern movement, until it passes a threshold that dictates: *Enough! We have enough information to make a decision. Press that button!* Some people have a high threshold—they are more cautious and require a lot of information before trying their luck at an action—and others have a low threshold—they are happy to sacrifice accuracy for speed.

All of these processes happen in the 650 milliseconds it takes me to form a simple perceptual judgment.

By investigating people's responses to dozens of these kinds of tasks, I discovered that there are parallels between how individuals form a perceptual judgment and how they make ideological judgments. There were two patterns in particular that elucidated how deep into the brain we could observe the reverberations of our political, religious, and social choices.

In such fast-paced perceptual tasks, participants are asked to maximize accuracy and speed. This creates a trade-off: either adopt a fast-and-furious approach, making rapid choices that inevitably entail some mistakes. Or go the slow-and-steady route, sacrificing speed but enhancing precision on each trial. We can quantify each person's *perceptual caution*—where they lie on the spectrum from fast-and-furious to slow-and-steady strategies.

The first insight is that individuals who were most cautious on these perceptual decision-making tasks were also the most politically conservative participants. In other words, perceptual caution in split-second decisions was correlated with political conservatism in policy preferences. Interestingly, conservatism is a linguistic synonym for caution. After all, what is conservative politics if not a politics that resists speed in favor of getting things right?

Caution does not mean slower reaction times in absolute terms. Instead it is about the balance between being fast and being accurate. Like typing on the keyboard or driving a car, some people favor going slowly and limiting their mistakes while others tend to go fast and let errors slip in. But there are some individuals who are (impressively) simultaneously fast and precise and those who are (heartbreakingly) both slow and error-prone.

Political conservatives' perceptual strategies tend to be tipped toward the slow-and-steady trade-off. We found that the conservative brain is cautious overall: cautious about morality and policy as well as in visual sensory decisions that last less than a second. Perceptual caution was evident in social and economic conservatives as well as in strong believers in nationalism, patriotism, the legitimacy of social dominance hierarchies and unequal status quos, and those who endorse violence and self-sacrifice for their ideological groups. This is consonant with other research showing that people with right-wing views tend to be more cautious when exploring new environments and are less likely to approach novel objects than people with left-wing beliefs.

The second insight from this study was that the individuals who exhibited a resistance to updating their beliefs in light of credible evidence were the slowest to integrate evidence and come to a decision in perceptual decision-making tasks. In other words, when we look at the evidence-accumulation rates, we find that individuals who are more dogmatic and resistant to evidence tend to integrate sensory evidence more slowly than their intellectually humble peers. The dogmatic mind's sensory evidence accumulation is inefficient and inflexible—it struggles to reliably extract high-quality information. Instead, the dogmatic mind experiences the sensory evidence as ambiguous and uncertain, even when it is not. When they learn and make decisions, dogmatic brains struggle to quickly weave together perceptual evidence—the colors, stimuli, feedback sounds, contingencies between stimuli and instructed responses. The dogmatic brain is slower to integrate the information into judgments. This difficulty in bringing sensory evidence together may hold clues about why the dogmatic individual struggles to integrate political evidence together in flexible ways.

This cognitive signature of dogmatism is markedly different from the perceptual caution that characterizes political conservatives. The distinction is important because conservatism and dogmatism are often conflated—yet they have different psychological origins.

This cognitive portrait of the dogmatic mind is notable in itself, but it becomes clearer when we examine some of the other findings in this data-driven study. We found that when we looked at the self-reported personalities of these dogmatic individuals, they did not profess to be slow thinkers, coming to decisions at glacial speeds. On the contrary, dogmatic participants claimed to be highly impulsive: they reported loving thrills and making rash choices.

And so the dogmatic mind may be one that makes premature and impulsive decisions based on evidence that was imperfectly understood. A dogmatic person's low-level unconscious cognitive machinery is slower, but their high-level self-conscious personalities mean they make impulsive decisions. This makes sense when we think of the profile of the dogmatic thinker—someone with a mind that resists updating their beliefs in light of credible evidence, someone who denounces ambiguity in favor of absolutes, who hastily rejects debate and prefers to ignore new or alternative information. If such a mind struggles to efficiently sort through diverse sensory evidence and find ways to unify it into a decision—if such a mind is a slow evidence accumulator but also tends to be imprudent, emotionally dysregulated, and prone to sudden choices—then it makes sense that the outcome will be a general dogmatism about any new inputs.

These findings suggest that the dogmatic mind's information-processing style is not restricted to dealing with ideological information. The dogmatic brain may have a more generalized cognitive and sensory impairment that manifests when the dogmatic individual evaluates *any* information, even at the speed of under a second. In these parallels between sensory perception and ideological intolerance, we see a domain generality and time invariance across different kinds of measurements.

Other researchers are uncovering convergent patterns showing that dogmatic and radical thinkers struggle to judge their own mental processes accurately—when performing cognitive tasks in which they must judge which of two black squares is filled with a greater number of flickering dots, ideologically radical individuals tend to be overconfident about the accuracy of their decisions. This implies that a dogmatic individual's mechanisms for judging themselves and interpreting the world are skewed for every kind of information—ideological and non-ideological alike. The dogmatic brain sticks to hastily formed interpretations, whether concerning political statements or bistable ducks and rabbits.

"Ambiguity—rabbit or duck?—is clearly the key to the whole

problem of image reading," wrote the art historian Ernst Gombrich in his seminal text on perception and pictorial representation, *Art and Illusion*. "Such an interpretation involves a tentative projection, a trial shot which transforms the image if it turns out to be a hit."

The act of interpretation—of making sense of an image or a piece of evidence or an ambiguous social situation—is an act that each of us approaches with a particular style. A style of interpretation that is sensitive to the subtleties of the object in front of us—attending to its vacancies and amorphous surfaces, recognizing its asymmetries and iridescent vibrations—or a style of interpretation that is more distant from the object, encountering it with preconceptions and snap judgments. The way in which our brains engage in interpretation is a reflection of how we approach the entire world.

In her famous 1964 essay "Against Interpretation," the scholar of art, history, and philosophy Susan Sontag argues that strong preconceived expectations about the meaning of an image or an idea interfere with honest perception. Interpretation is a process of layering preconceptions on top of our experience in order to figure out what a thing—a painting, a text, an appearance—really means.

For Sontag, this interpretation process—this translation into meaning—is deeply fraught. "The modern style of interpretation excavates, and as it excavates, destroys; it digs 'behind' the text, to find a sub-text which is the true one," Sontag writes. "The most celebrated and influential modern doctrines, those of Marx and Freud, actually amount to elaborate systems of hermeneutics, aggressive and impious theories of interpretation." An interpreter—a critic, a preacher, a prophet, a psychoanalyst, a leader—claims to disclose true meanings and, in so doing, estranges us from the world as it can be experienced by our senses. "To interpret is to impoverish," Sontag suggests, "to deplete the world—in order to set up a shadow world of 'meanings.'"

Sontag argues that imbuing art—texts, paintings, photographs, cinema—with interpretation is an act of reduction, of turning our eyes away from "the luminousness of the thing in itself, of things being what they are."

In art—as in other matters of faith—interpretations can be akin to ideologies and ideologies can impose interpretations. If one adheres to a doctrine rigidly, every perceptual experience becomes subjugated to meanings that fit the doctrine. Ideological structures and significances lead to an impoverishment of our sensory world. Rather than experiencing art or the world directly, we come to experience it indirectly, inauthentically, ideologically, avoiding its ambiguity in favor of predetermined meanings.

To cut through these stifling interpretations, we need less ideology and more direct sensation. "What is important now is to recover our senses," Sontag proposes: "We must learn to *see* more, to *hear* more, to *feel* more." In Sontag's words, emphasis should be placed less on interpretation and more on intercourse. Less on logic and more on erotics. Uncensored by interpretations that structure our perception, we must return to the body, back to our physicality, back to unfiltered sensation.

For Sontag, interpretations constitute the "the revenge of the intellect upon the world." For us, feeling more, sensing more, can become our revenge on ideology.

15

YOUR EMOTIONAL FINGERTIPS

It is in the faintest of gestures—the imperceptible stirring or uncontrolled twitch—that we disclose the grandest emotions. Within our bodies, under our skins, there is relentless, jazzy motion. Muscles tense and relax in spasmodic melodies. The brain, gut, and lungs chatter in breathless conversation. Blood is pushed and pulled through webbed arteries, capillaries, and veins. Nerve cells fire in synchrony, forming steady cyclical beats with every brainwave that swells and collapses and rises again. Surprises and prediction errors provoke series of tangoed syncopations.

This inner movement—invisible to the outside world, sometimes invisible to ourselves—creates rhythms and tempos, symphonies inside each body. The visceral vibrations that ripple with every mood are musical markers that echo and sustain our body's balance. Our biochemical homeostasis.

The sounds of our organs betray our innermost emotions. To come into awareness of our panic or passion, we peg our hands to our chests, feel the acoustics of our tell-all heartbeats or the cadence of our breath. In moments of uncertainty, we ask the doctor to listen in too. With a ropy stethoscope, the doctor attends to the rumbling ocean beneath our skins. *Listen,* we beg, *what can you hear? What can you tell me about my body?*

These are private, introverted resonances, revealing our feelings even before they diffuse into consciousness. "Feelings let us *mind the body,*" notes the neuroscientist Antonio Damasio; "feelings offer us a glimpse of what goes on in our flesh."

Yet when we describe political emotions or feelings of faith, when we imagine the emotional ideologue, we expect an extroverted theatricality. Foaming rage or rapture. A boiling heat. Emotions visible on every face, in every ribbon of propaganda, within every mob that has been whipped up into the stiff peaks of violent fury. These political emotions are paraded with exaggerated clarity—*Fear! Disgust! Anger! Hate! Pride!*—as though the ideologue is an exaggeration in itself, an exaggerated self. But not all feelings are melodramatic or conspicuous. Many of the signatures of our emotions are slight, implicit, faster than our rationalizations or excuses. These ephemeral indicators are also more attuned to our ideological worldviews than we might dare to imagine.

In 2008 an experiment was published in the journal *Science* that introduced the world to the possibilities of political psychophysiology, the study of how physiological processes are altered by political processes. Nebraskan researchers exposed participants to abrupt noises—*BOO! BAM! CLAP!*—and measured the size of their startled blinks. With a couple electrodes placed below the lower eyelids on the orbicularis oculi muscle, they could differentiate hard blinks from gentler flutterings. Do we all respond equally to sudden and frightening sounds? Seemingly not. The researchers selected participants from the political spectrum's outer edges—the most ideologically extreme, either to the political left or to the right. The results demonstrated that individuals who were more politically conservative tended to blink more tightly in response to the threatening bursts of noise. In contrast, liberals reacted less strongly, blinking more softly, betraying less alarm, less instinctual fear.

The effects were present not only for auditory sounds but also for visual stimuli. When presented with fear-inducing images, such as a photo of a huge spider resting ominously on the face of a terrified person, or of a person who has been wounded, the researchers found that conservatives viewing threatening images were more physiologically aggravated than liberals. By taping electrodes to participants' fingers, the researchers could measure their skin conductance response, an index of slight changes in the skin's electrical activity that mark the biological cascades of physiological arousal. These short-term fluctuations on our fingertips are triggered by heightened sweat secretion in the eccrine glands. This leads to better conductance of electricity through the skin, which can be measured by passing a weak and painless electrical current through two electrodes on our fingers or feet.

These physiological markers become activated when the human body encounters a stressful, surprising, or novel situation. In these moments, the sympathetic nervous system gears into action and skin conductance rises. The sympathetic nervous system is the branch of the autonomic nervous system responsible for the infamous fight-or-flight response. When the body perceives danger, it is prompted to accelerate the heartbeat, quicken the breath, tighten the muscles in preparation for action, and precipitate tiny beads of sweat along our brow and in the creases of our palms. By measuring sympathetic nervous system activity through the skin conductance response, neuroscientists can reverse engineer our anxieties and arousals, inferring how strongly a nervous system is stimulated by an event. These physiological signals transpire regardless of whether the person is aware of their feelings, before they are even conscious of their emotions or can give them shape through language.

The finding that there are ideological differences in how startled and alarmed people are in response to frightening noises and images suggests that our bodies are entangled with ideologies in far-reaching ways. The researchers must have known the delicate and controversial ground they were treading. In the paper describing the results, the authors labeled conservative political issues such as support for school prayer and opposition to gay rights and abortion as "socially protective" policies rather than the more standard descriptor "conservative" or "right-wing" policies. (Incidentally, the academic team included more authors named John than authors who were women, but that's a separate issue.)

Statistically, the study was on shaky ground. The sample was small—only forty-six participants—and the difference between the physiological reactions of the conservative and liberal groups only just passed the threshold for statistical significance. Conceptually, however, the study was inventive. It said *something*, something deeply interesting and provocative.

The results had a seismic effect on the field. Multiple international research teams began chasing the elusive physiological marker of political ideology. Within five years of the *Science* paper's publication, an entire journal issue was dedicated to the debate about whether conservatives' ideological preferences are undergirded by a psychophysiological *negativity bias*, a heightened biological responsivity to negative information relative to positive or neutral information. Across dozens of densely printed pages, more than fifty leading scholars debated the validity and nuances of this hypothesis, asking whether and in what ways conservatives and liberals differ in their physiological sensitivities.

How gracefully has this debate aged since 2008? There is both evidence in favor of the negativity-bias hypothesis and evidence against. The crux of the question lies in whether there is a *generalized* negativity bias that conservatives exhibit in response to any stimulus or situation they encounter, not simply in response to political news. A growing collection of experiments supports the hypothesis. For instance, one experiment found that when encountering foggy images of faces that have an ambiguous emotional expression, conservatives were more likely to interpret them as threatening. Other experiments found that when conservatives were presented with negative and positive perceptual information, their attention was more strongly attracted by the negatively tinged information.

At the same time, some research groups that have tried to directly replicate the 2008 study have failed to find either a consistent threat sensitivity dimension or a link between threat sensitivity and conservatism. Many others are hesitant to let the negativity-bias hypothesis go. Other critics question the focus of such investigations and wonder why this research seeks to explain why a person would be conservative rather than why a person would be liberal. Some accuse the politics of academia, arguing that left-wing biases creep into empirical undertakings. A more charitable interpretation frames the negativity-bias hypothesis as an exercise in empathy: liberal thinkers trying to decipher the anxieties and sensitivities that spur individuals toward conservative worldviews.

Some suggest that it might not be sensitivity to *threat* or *negativity* that differs between political conservatives and liberals, rather that it is a more specific sensitivity to *disgust* that motivates individuals to seek out conservative ideologies. In this formulation, disgust sensitivity drives individuals to adopt traditional purity-oriented moralities and to feel hostility toward any issue that seems to invoke themes of bodily or sexual impurity or transgression, such as abortion, gay rights, and immigration. *Squeamish about the body? You may be squeamish about politics too.* This parallels legal philosopher Martha Nussbaum's proposal that "disgust has been used throughout history to exclude and marginalize groups or people who come to embody the dominant group's fear and loathing of its own animality and mortality."

Experimental paradigms have emerged that measure people's natural responses to potentially disgusting stimuli. Some experiments use self-reported questionnaires of disgust sensitivity and others use physiological measures to quantify people's skin conductance response when they view disgusting images. These experiments have found that there is a consistent correlation between conservative policy preferences and disgust reactions to scenes of contamination, such as images of infection symptoms and disease. For instance, in one study with thousands of Danish and American participants, individuals high in sensitivity to disgust were more likely to support anti-immigration policies. A person's degree of disgust sensitivity was linked to their disapproval of close contact with immigrants, such as having immigrant families as neighbors or being willing to swim in public pools that are frequented by a large number of immigrants. *Scared of physical contamination? You might be scared of the illusion of political or racial contamination too.* The relationship between disgust sensitivity and anti-immigration attitudes was present for both conservatives and liberals. In fact, the relationship was most pronounced among self-reported liberals. Individuals who are both left-wing and highly sensitive to contamination disgust are likely to be antagonistic to close contact with immigrants, which can lead them to be at odds with the touted policies of left-leaning political parties.

In all of these studies, the assumption is that traits regarding reactivity to negative, threatening, or disgusting information in the nonpolitical sphere shape the ideological worldviews people come to hold. If you are easily threatened, you may gravitate toward a conservative ideology that seeks to buffer against, or at least obsessively explain, feelings of threat. If you are prone to feeling trust rather than disgust, you will be more open to accepting minorities who deviate from local traditional norms.

These theories and experiments are both fascinating and troubling.

It is important to remember that these findings are general patterns and correlations, not determined outcomes. Not every person who is sensitive to disgust will be predisposed to electing politicians who spew rhetoric against gendered or racialized minorities. Other psychological traits—such as cognitive flexibility or emotional regulation—can intervene or subdue these tendencies. After all, each individual is an amalgamation of dispositions and sensitivities.

What are we talking about when we talk of *sensitivity*? The object of our sensitivity is what distracts and destabilizes us in some way. Sensitivity is not fragility. To be sensitive to something is to struggle to ignore it, to see it rather than overlook it, to feel it intensely. The French word *sens* captures the nuances of this phenomenon: it is sensory-sense, significance, attitude, and direction. If we are sensitive to the weather, we feel acutely its drying effect on our skin or the way it alters the density of the air we inhale. If we are sensitive to imminent threats or intrusions into our private space, we will notice them on our horizon and experience a kind of energy surge in our bodies. Our perception of the world will be endowed with a readiness to be riveted or moved. Sensitivity is a kind of alertness: alertness both in the sense of perceptive awareness and in the sense of alertness to potential alarm.

A particular sensitivity is an intuitive, automatic, and rapid orientation toward something—similar to what the phenomenologist philosophers thought of as a consciousness of something. As phenomenology's founder Edmund Husserl observed about our experience of the world:

> this world is there for me not only as a world of mere things, but also with the same immediacy as a *world of objects with values, a world of goods, a practical world*. I find the physical things in front of me furnished not only with merely material determinations but also with value-characteristics, as beautiful and ugly, pleasant and unpleasant, agreeable and disagreeable.

When we measure an individual's sensitivity, we glean their instinctual associations and reactions—what they consider ugly and what they find mesmerizing, what rouses their sympathy and what leaves them unmoved. What they value in the world.

The lesson of political psychophysiology is *not* that conservatives are physiologically sensitive and liberals are physiologically indifferent *at all times for all things.* Sensitivity is not a single trait; we have a multitude of sensitivities. Because sensitivity is always oriented *toward* something, sensitivity is always *of* a kind of stimulus. What we care about matters. Some domains will spark the liberal body more than the conservative body, and vice versa. Other issues will trigger the nervous systems of extreme ideologues more than moderate believers. The question becomes *who* is most sensitive to *what.* And why these sensitivities matter.

Recent experiments have begun to look for relationships between ideology and pain sensitivity: *How easily do you feel pain? How much pain can you tolerate? Could your sensitivity to pain shape your empathy to others' concerns?* Some researchers search for links between ideology and taste sensitivity: *Is your sensitivity to bitter tastes a predictor of your bitter morality? Is ideology a matter of taste? Stick out your tongue and let us check!* Other groups have begun to investigate interoceptive sensitivity, the degree to which a person is attuned to their own internal bodily states. Interoceptive sensitivity is measured by how accurately a person estimates the rhythms of their own heartbeat when sitting still and focused. In interoception experiments, participants' pulses are monitored while they listen to a soundtrack of auditory tones that are, in some conditions, synchronous with their heart rate—participants are basically hearing their own heartbeats played back to them—and, in other conditions, out of sync with their beating pulse. *Could your sensitivity to your own internal bodily signals be linked to your willingness to numb your own discomforts and be dominated by an ideological cause instead?*

In all these weird and wonderful ways, scientists are leveraging the body's internal music to decipher its sensitivities and study whether they predict its ideological sympathies. The hypothesis is that ideologies are truly embodied phenomena: our physical senses, physiological reactions, and biological experiences sculpt and are sculpted by our ideologies.

Some researchers have tackled the question of ideological differences in physiological sensitivities from another angle. Rather than presenting nonideological stimuli such as creeping spiders or aversive images of sickness and disease, some political psychologists have studied the physiological arousal of individuals while they are watching explicitly politicized information, such as videos on immigration, wealth redistribution, or climate change. Experiments that take this approach have shown that individuals with more extreme attitudes—regardless of whether the individual leans to the political left or political right—will have heightened physiological arousal to videos about politicized topics, as measured by the skin conductance response. Watching a video about refugees crossing seas to enter the participant's country, the ideologue's body reacts strongly. Moderate individuals' bodies mimic their moderate, temperate attitudes. Their physiological response to polarizing news is more subdued.

In another investigation by political psychologists in Amsterdam in 2023, participants viewed angry, happy, or neutral emotional expressions of political leaders from the Dutch parliament while undergoing facial electromyography, which recorded the activity of the muscle region above the eyebrow responsible for frowning (called the corrugator supercilii muscle) and the activity of the muscle region responsible for smiling (the zygomaticus major muscle). Participants frowned a lot more when they viewed political leaders they disliked—*who doesn't grimace at the detested opposition?*—except that when participants saw that the outgroup leaders were angry, there was also activity in the smiling muscle—*happy to see the anger of the opposition!*

"Our emotional lives are very much informed by ideology," recognized the political philosopher and activist Angela Davis. "We ourselves often do the work of the state in and through our interior lives." Ideologies can structure our interior emotional lives in ways that can be heard across all levels of the body: from the conscious emotions to the unconscious feelings to their physiological markers. Angela Davis is a titan of the civil rights movement, an astute observer of injustices, and also a proud and bold proponent of the idea that violence can be a legitimate form of resistance. Davis's writings suggest that oppressive ideologies and the state teach us to tolerate inequalities that we should find abominable. In her body of work, Angela Davis conveys that "bad" ideologies justify inequality and that "good" ideologies are driven to quash such hierarchies. If we wish to understand the politicization of our bodies, we need to see how ideologies modulate our sensitivities to inequality.

New research addresses this question, examining how our attentional and physiological responses to sights of inequality are interlinked with our ideologies about inequality. Experiments have found that people with more egalitarian beliefs are more attentive to signs of inequality against marginalized groups (women, racialized minorities, unhoused people) in images of urban scenes or numerical graphics. This sensitivity is not a bias toward seeing inequality everywhere. Egalitarians will report evidence of inequality only when it is present. The sensitivity to inequality means that when there are true disparities, egalitarians will detect them. People who justify hierarchies will simply not notice the inequalities, even when the researchers offer them financial incentives for delivering accurate answers.

Bringing this into physiology, researchers looked at individual differences in how troubled people are by economic disparities. Participants' levels of system justification were measured—whether they justified economic capitalism as fair and legitimate or they saw it as a source of extreme and unfair inequity. Participants were then exposed to videos of interviews with people experiencing homelessness, who discussed their daily routines and the adversities of living in poverty. Subsequently, participants watched "control" videos of people describing fishing or coffee making. The physiological responses that were elicited while watching the homelessness videos were contrasted with those that took place while watching the neutral control videos.

People who rejected the validity of stark economic inequalities had markedly higher negative arousal while watching the videos describing homeless adversities than when they watched the neutral videos. Their bodies revealed their distress. In contrast, for system-justifiers, the psychophysiological markers did not spike in response to the homelessness videos. In fact, their physiological responses to the adversities of homelessness were indistinguishable from their physiological responses to videos about coffee making or fishing. Their bodies expressed little anguish, sadness, anger, or distress at witnessing another's suffering. The ideologue believing in hierarchies is viscerally numb.

Our most private physiological responses betray our ideologies.

If ideological echoes can be detected in the peripheral nervous system—the extension of the nervous system outside the brain and spinal cord that carries signals into the rest of the body, including our heart and fingertips and dainty eyelids—what can we see when we look inside our heads, at the central nervous system? What happens to our brains when we internalize dogmatic belief systems?

16

AN IDEOLOGY WALKS INTO A BRAIN SCANNER

The act of scanning a brain is spurred initially by curiosity, by a scientific question. But it is also an act of medical measurement, of enforcing transparency on a closed structure. It is an exercise in exposing matter to light, deconstructing a tightly bound construction, generating a most intimate photography. Perhaps it has the flavor of voyeurism.

With MRI's translucent magic, it becomes possible to examine the grayscale X-rays of all that furnishes the space behind the human face. White matter corridors, bumpy blobs, air-filled sacs, and lumpy globules. The thick bones encasing the human head become luxuriously sheer and diaphanous.

The political neuroscientist can then search for consistent patterns between ideologues of different persuasions. What are the contours and interiors of an ideological brain? Will people with different ideologies have brains with subtle differences in structure and function? Will the brain of the compassionate egalitarian diverge from that of the old-fashioned authoritarian? How will the cerebral foldings that form the brain's valleyed surface vary between the moderate and the fundamentalist? Will the ridges of peaking gyri and the passes left by sunk sulci possess different indentations for the right-wing believer relative to the left-wing believer? And if so, what do these neurological differences mean?

The challenge of political neuroscience is not a problem of dealing with a paucity of studies that must be stretched or boring results that must be transformed into matters of relevance. On the contrary. New studies are published every month, each more dazzling and beguiling than the last. The challenge is to interpret the results in a balanced way that avoids sensationalism. It is easy to claim that a certain network of brain regions is activated every time we observe a detested political leader or pray softly on our religious revelations. *Of course* the brain is activated. Every thought is a biological mark, and so the mere presence of a neural activation means little except that we are alive and conscious organisms (in itself great news, but of limited use to the political neuroscientist).

Several recent studies have illustrated that when participants watch inflammatory political videos, the brains of left-wing participants are "synchronized" with those of other left-wing participants while the brains of right-wing participants are "synchronized" with those of other right-wing participants. *Neural polarization*, scream the headlines. But all it is likely to mean is that similarly minded people respond to things in similar ways. Simply because a brain region is activated while a participant is gazing at the image of their most hated politician reveals, potentially, very little. Other studies employing machine-learning techniques have shown that it is possible to infer participants' political conservatism from the brain activity they exhibit while inspecting a nonpolitical image of a disgust-linked physical injury. Sounds ominous, potentially exciting. But what does it mean? Maybe that, as the negativity-bias hypothesis suggests, conservatives and liberals have different fears and discomforts; or that machine-learning algorithms can extract a lot from minimal data.

So when does a neural pattern tell us something compelling about the nature of ideological thinking? When is a neuroscience of ideology a promising field and when is it a futile exercise?

Warnings and cautious qualifiers are rarely riveting (limits on our imagination seldom are), but they are intellectually honest. To learn the most from political neuroscience, it is better to embrace a critical and attentive outlook instead of a grand persuasive sweep. With every experiment, let us ask what *ideological domain* is being explored—is the study focusing on political identities or radicalization or religion? Which *brain regions* are implicated—and what, if anything, can this tell us about the theories of ideology that are developed as a result? Do we discover the involvement of brain areas responsible for analytical decision-making or regions for emotional processing or areas commonly associated with entirely different functions? Is the focus on brain *structure* or brain *function*? Do we know much about the *participants*—the sample size; the participants' citizenship, geographical location, or ideological affiliation; or whether they represent a diverse community as opposed to a homogeneous and privileged student pool? Are the researchers asking participants to perform a certain *experimental task* in the brain scanner or are the participants lying delightfully still, eyes closed, mind wandering? How substantial and significant are the *reported effects*? Does the *theoretical interpretation* of the results seem valid or inflated or too uncertain to tell? Is there a question about *mechanism* or simply a search for an effect?

These are the kinds of questions that political neuroscientists and their students are constantly grappling with, and I invite you to imagine wrestling with them too. You may detect the difficulties in making sense out of a field that is new, emerging, and still under development. You will notice experimental designs marked by care and some characterized by clutter. The finest neuroscientific studies are built up slowly, iteratively, and thoughtfully. When learning about them, it is valuable to adopt the same caring and critical approach.

One of the first studies that placed individuals in the brain scanner and searched for structural differences according to their ideology took place in 2011 and focused on political beliefs. (Bizarrely, the research team included the British actor Colin Firth, who had guest edited a BBC Radio 4 show about the nature of politicians' brains.) Would the liberal brain be differently sized and proportioned relative to the conservative brain? A team of London-based researchers found that more conservative people tended to have a larger right amygdala than political liberals.

The amygdala is famously an almond-shaped structure that governs the processing of emotions, especially negatively tinged emotions such as fear, anger, disgust, danger, and threat. Because we often hear about *the* amygdala, most people assume that we have a single amygdala-almond tucked inside the middle of the brain. But in fact it is a paired structure of two parts. We have two amygdalae, one in each hemisphere of the brain. The London researchers found that the right amygdala (but not the left) was enlarged in more conservative participants. This pattern was corroborated in another sample of additional participants at the time, and it has since been replicated in a much larger and more diverse sample of over 900 participants by an independent research team in Amsterdam.

In a separate study by researchers in New York, amygdala size was found to predict individuals' level of system justification, which reflects the degree to which people support unequal social systems and prefer to maintain the status quo. System justification is strongly correlated with conservatism, as both refer to adherence to past traditions at the expense of change. Yet system justification and conservatism are dissociable constructs. Individuals who are high in system justification regard the existing social system as legitimate and desirable, even if it reinforces existing inequalities. A strong system-justifier not only will accept inequalities as legitimate but may also promote them as necessary. This is not always a matter of self-interest: system-justifiers will sometimes support inequalities that are detrimental to their own welfare. In two independent neuroimaging experiments, the New York team found that system-justifiers were likely to have greater combined left and right amygdala volume on average.

Since the amygdalae store emotional associations about threat, fear, and disgust, along with learned information about social hierarchies and relationships of dominance, political neuroscientists have interpreted these findings as reflecting a natural affinity between the function of the amygdalae and the function of conservative ideologies. Both revolve around vigilant reactions to threats and the fear of being overpowered.

But *why* is the amygdala larger in conservatives? Is it because conservatives are overreactive to negative information and this overreactivity translates into a bigger amygdala? In general, the size of a brain region is linked to its processing capacity, but the degree to which structural anatomy depends on and responds to functional activity is still debated among scientists.

The ambiguity around these results perfectly reflects the stubbornness of the chicken-and-egg conundrum. Do individuals with larger amygdalae gravitate toward more conservative ideologies because their amygdalae are already structured in a way that is more receptive to the negative emotions that conservatism elicits? Or is the experience of being immersed in conservative, system-justifying ideologies an experience that alters our emotional biochemistry in a way that leads to structural brain changes?

The question of causality lingers. Disentangling which way the arrows point is an ongoing endeavor.

From the amygdala, let's travel up the pathways of the *limbic system*, the neural system named for its edgedness (*limbus* is Latin for "edge"). This system is a circuit of brain regions that communicate with each other through anatomical tracts that facilitate functional connections along the borders between the frontal cortex and deeper midbrain structures. Through the limbic system, human beings process emotion, uncertainty, and the values of rewards and punishments. If we follow the biological projections of the amygdala up and outward, we encounter the next stop in the limbic system: the anterior cingulate cortex, abbreviated as the ACC. The anterior cingulate cortex is a crescent-shaped region that hugs the corpus callosum—the bridge between the brain's two hemispheres—and sits snugly between the outer layers of the frontal cortex and the rest of the limbic system. The ACC is a long sausage-like structure that is functionally diverse and characterized by a gradient of subdivisions involved in emotional processing and cognitive control. Tellingly, there are no discrete borders between where "emotional processing" ends and "cognitive processing" begins; rather, there is a *gradual* change in function within the anterior cingulate cortex. In our concepts and in our anatomy, the traditional division between cool intellect and heated emotion is a mirage that ought to be dissolved. The neural mechanisms of affective and rational processes overlap and share anatomical spaces.

Not only is the ACC an impressively multidisciplinary organ, it is also a hub that has unusually high connectivity to other parts of the brain, particularly the rest of the frontal cortex. It is therefore thought to be a key coordinator of complex cognition.

It may not be entirely surprising that when neuroscientists study the processes underlying political and religious beliefs—these conglomerates of dogma and passion—the anterior cingulate cortex frequently emerges as a focal culprit. When the London-based researchers looked for links between political ideology and brain anatomy, they observed that more liberal participants had a larger anterior cingulate cortex. This finding emerged in two independent pools of participants in London. But disappointingly for the anterior cingulate cortex vying for the political neuroscientist's attention, the Amsterdam and New York research teams struggled to replicate this effect.

The inconsistencies in the reproducibility of the ACC's role in ideological thinking may be a result of overemphasizing the importance of the size of brain regions, when it can be more instructive to look directly at their function instead. And when it comes to connecting brain function to ideological thinking, the ACC flares into a luminous rouge—a brain region haughtily aware of its own importance.

The ACC monitors the errors and conflicts that arise when a person processes information. It generates a signal that is akin to a little alarm bell that rings every time an error is made. For some individuals, the bell rings loudly—signaling that a mistake has been committed—while for other individuals, the bell is more muted—blunders are barely registered. The ACC is therefore crucial for behavioral adaptability and for recognizing when a habitual action is no longer appropriate and must be replaced with a new tactic.

Neuroscientists study the ACC's error-monitoring functions by inviting individuals to complete mental inhibition tests called Go/No-Go tasks. In an inhibition task, a participant learns a habitual action, called a *Go* action, such as pressing a button every time a green circle pops up on the screen. Infrequently they encounter a *No-Go* stimulus, such as a red cross that instructs them *not* to make the action the next time they see a Go signal, to halt and withhold the habitual act. A person who inhibits well makes few Go actions when they have been shown the No-Go signal. *Attentively, they monitor the traffic lights.* A person who struggles with inhibition will impulsively continue to Go even when the No-Go red light has flashed. *Ignoring the flaming red traffic light and hoping for the best.*

When we dot participants' scalps with electroencephalography (EEG) electrodes and ask them to complete this inhibition task, the ACC produces reliable signals, electrical changes called event-related potentials. Event-related potentials are voltage changes reflecting concentrated neuronal firing, when neurons all respond to a psychological event in concert. If an individual commits an error, we can capture a signal called the error-related negativity (ERN), which is the brain wave that emerges around 50–100 milliseconds after an error has been made. Each person's ERN can be measured to capture the degree to which their brain senses a conflict between the habit they have learned and the need to inhibit it.

Across multiple studies, self-reported political liberals and egalitarians were found to have greater ERN amplitudes, meaning that the brains of liberally minded individuals were more sensitive to errors and conflicts in inhibition tasks. In contrast, political conservatives' ACC emitted a weaker ERN signal, reflecting a more dulled reaction to their own errors. Larger error-related negativity, as found in liberal participants, is linked to better inhibition; the more you are aware of your errors and the need to overcome habitual responses, the better you are at avoiding preestablished habits when these are wrong or maladaptive.

Converging patterns between ideology and the ERN are also found in the study of religion. Individuals with low religious zeal and weaker belief in God tend to have greater error-related negativity in neutral cognitive inhibition tasks, similar to political liberals. Stronger attachment to religious dogma is linked to diminished sensitivity to cues that extinguish habitual behavior. One experiment with Mormon students in Utah showed that when religious believers are prompted to think about God's love and mercy, their error-related negativity declines further. This suggests that reflecting on God's unconditional love diminishes the believer's sensitivity to errors and conflict. Imagining the forgiveness of divine protectors might render us less attentive to our own mistakes. The relief of redemption runs deep.

These instances of political neuroscience exemplify the field at its most ambitious: seeking to unearth neural processes that are invisible to outside observers but which elucidate the far-reaching implications of ideologies on brain function.

As we travel upward from the amygdala and the ACC to the frontal lobes, onward through the limbic system, we journey further into the frontal cortex and enter the most widely recognized and celebrated area of the brain: the prefrontal cortex. It is the crown behind our forehead, the seat of our most complex decision-making, a set of interconnected hubs that govern high-level mental computations and transform them into conscious decisions. When we place our forehead in our hands, it is the prefrontal cortex we are comforting. When we invent narratives or formulate sophisticated criticisms about the book in our laps, it is the prefrontal cortex we are engaging. When evolutionary psychologists gloat all too arrogantly about the superior rationality of human beings, it is the prefrontal cortex they point to.

In what ways does the prefrontal cortex shape our ideological commitments? A first clue emerges using one of the oldest methods in the psychology handbook: naturally occurring brain lesions.

One study used a unique dataset of patients with brain lesions to study the ideological leanings of individuals who experienced injury, surgical intervention, or atrophy of the prefrontal cortex and compared them with healthy controls and patients with lesions of the anterior temporal lobe, where the amygdala is located. Patients with frontal lesions were more politically conservative than healthy controls or patients with lesions in the anterior temporal lobe. And the greater the damage to the dorsolateral prefrontal cortex—the surface region of the prefrontal cortex above each eye, close to our hairline—the more conservative the patient.

This was not true of amygdala damage: the percentage of damage to the amygdala was not linked to patients' political conservatism or liberalism. If you were hoping to reverse conservatism by shrinking the amygdala, or show that damage to the brain's emotional core is enough to disrupt ideological identities, you are in for disappointment.

The dorsolateral prefrontal cortex is linked to levels of conservatism and also to more general radicalism. Studies of former American soldiers who were combatants in the Vietnam War and suffered penetrating traumatic brain injury shed light on these processes. Damage to the dorsolateral and ventromedial prefrontal cortex—the inner region that hides behind the space between our eyebrows—was found to be linked to political radicalism and religious fundamentalism. Brains with lesions to the ventromedial prefrontal cortex judged radical statements as moderate, whereas brains with no lesions or lesions elsewhere accurately discerned radicalism and condemned it. This signifies that damage to the ventromedial prefrontal cortex is linked to perceiving extreme actions and policies as morally permissible. War veterans with injuries to the ventromedial prefrontal cortex and dorsolateral prefrontal cortex also held the most religious fundamentalist views about their Christian faith. Notably, the damage to the dorsolateral prefrontal cortex was linked to reduced cognitive flexibility on a task similar to the Wisconsin Card Sorting Test. This diminished flexibility in turn predicted heightened fundamentalism. By rendering people more cognitively rigid, the prefrontal brain injuries led to greater religious fundamentalism and radicalism.

It may be tempting to conclude that since the prefrontal cortex's integrity is crucial for the possession of liberal beliefs, then to be undogmatic and irreligious and progressive is to have a large intact prefrontal cortex. (All leftists who yearn for neuroscience to validate the rationality of their belief systems: applaud and rejoice!) *Not so fast.* The prefrontal cortex is a vast, dense, and convoluted area with many subdivisions that are recruited into different neural circuits, webs, loops, and projections, which are governed by different neurotransmitters, hormones, and enzymes.

Rather than a contained region with a singular job, the prefrontal cortex is akin to an international transportation hub: a network of crisscrossing connections. It has an airport with multiple terminals as well as railway connections, bus services, and routes for ushering travelers from one circuit to another. The terminals of the prefrontal cortex transfer signals back and forth to the amygdala and anterior cingulate cortex as well as their neighbors, the memory-storing hippocampus and hormonal hypothalamus. The prefrontal cortex is in constant communication with the movement-coordinating cerebellum and the dopamine-rich striatal system for learning. Both the dorsolateral and the ventromedial sections of the prefrontal cortex whisper back and forth to the insula, the organ monitoring our subjective feelings and interoceptive states that create an insulated sense of "self." There are reciprocal connections between the prefrontal cortex and other regions, with inputs flowing in both directions within a closed loop, as well as nonreciprocal connections, where projections flow away from organs but not back. Through this lattice of overlapping connections and loops, the prefrontal cortex binds sensory impressions, memories, and predictions into conscious behavior. Thinking of the prefrontal cortex in terms of the transfer, flow, and manipulation of information across global brain dynamics is more informative than envisioning it as an isolated cognitive overlord that alone dictates our sophisticated behavior. Brain mapping is not an isolationist task, carving up the cortex into discrete units. Instead, connectomics is the name of the game.

So a political neuroscientist hoping to learn how the prefrontal cortex's myriad operations facilitate ideological thinking may need to zoom into live dynamic processes rather than overemphasize static differences between groups. One way to do this is to conduct a deep dive into each person. Rather than investigating interpersonal differences, study intrapersonal phenomena. Look at the neural activity that transpires when a brain is thinking ideologically, dogmatically, extremely, and compare this to when the same brain is thinking more soberly, more flexibly. By studying the moment-by-moment neural activations that take place while people evaluate ideological narratives, policies, and values, we can gather a sense of how the different prefrontal regions and networks scaffold ideological reasoning.

Sacred values sit at the apex of ideological thinking. These are the values for which we are willing to fight and die—the values to which we feel duty-bound and existentially tied. In experiments by researchers in Barcelona, brain activity was monitored when ideologues contemplated their most cherished sacred values, such as their beliefs in God, their readiness for armed conflict, or their opposition to same-sex unions. The neural patterns for sacred values were then compared with the neural activations for nonsacred values—values held less dogmatically, values that are negotiable rather than those that are inviolable and sacrosanct. A nonsacred value is a value we would be willing to compromise on. *Would you accept money to give up your ideological cause? Could you accept concessions?* Nonsacred values inspire less fervor to violence and self-sacrifice. They are potentially disposable. Sacred values, on the other hand, are deemed as nonsubstitutable. No material improvement or social concession can replace or diminish sacred values' potency. Sacred values feel absolute and divine.

In experiments with far-right Spanish citizens and with Muslim immigrants who endorse militant jihadist causes, the Barcelona research team explored the neural signature of sacred values. Subdivisions of the prefrontal cortex, such as the ventromedial prefrontal cortex and the inferior frontal gyrus, were uniquely involved in the processing of sacred values but not in the evaluation of nonsacred values. Moreover, the greater a militant jihadist felt identity fusion with Islam and the Muslim Ummah, the less there was recruitment of the dorsolateral prefrontal cortex when they evaluated sacred values. Our strongest commitments uniquely modulate neural processes that take place across the prefrontal cortex.

But it is not only the prefrontal cortex that is implicated. Functional connectivity between frontal areas and attentional areas becomes more salient when a person reflects on sacred values. When ideologues think about the sacred values for which they are willing to die, there is heightened activity in the nearby anterior cingulate cortex as well as the insula responsible for our sense of self. Mentalizing networks involved in empathizing with others, such as the temporoparietal junction, also get activated when participants consider their willingness to commit violence for their ideological values. It is not a single region that lights up—the whole brain is ignited, from the prefrontal cortex's outer edges to the core. Thinking ideologically, thinking about the sacred, thinking about who we are willing to harm and who we are determined to protect, shifts our neural processes in specific ways.

The longer we look at the prefrontal cortex and extremist beliefs, the more we are forced to look more broadly at the rest of the brain. And the further inward we trace the grooves of this captain of an organ, the more we begin to ask again about how causes and consequences feed each other, about how we should understand the tension between biology and change. Biological accounts of ideologies do not imply that individuals' views are fixed and unchangeable. On the contrary, decoding the neural representations of our dogmas can help us see the ways in which our brains and our values are malleable and ready for transformation.

Part V

FREEDOM

17

SPIRALING IN AND OUT

Ideologies are tales of inevitability. Logical laws, commandments and prohibitions, dystopian dangers and utopic dreams, all work at the service of a causal story that exudes an ambience of necessity. In these narratives of predestination—whether religious or romantic, nationalistic or with an aura of the scientific—there is a sense that something is "meant to be" and "cannot be otherwise."

Determinism hums at the core of ideologies. Within determinist frameworks, free will is deemed a naive illusion or, at least, a dangerous attitude. Each individual's future is primarily controlled by the grand narrative of its past. Choosing a future is futile because it has already been chosen for you. The only permissible kind of agency is the agency to follow the ideology's prescriptions and enact its vision of paradise. Trying to defy the ideology is to tempt the gods' wrath—to invite a sinister and inevitable catastrophe.

The story of the ideological brain is not one of inevitability. It is a story of potentialities that get expressed and suppressed. We all lie on a spectrum of suggestibility. This spectrum reflects the interaction of all the traits that make people susceptible to ideological thinking: their biological or cognitive dispositions, their personalities and social experiences, their traumas and perceived resources, abundances, realities, and absences.

Spectrums are less neat than clear binaries. It is tempting to imagine that there are two categories of people: the vulnerable and the resilient. The radical minority who can be manipulated into monstrosity and the sensible majority who are immune to dangerous extremist ideologies. (Or the reverse—the sensible minority and radical majority—depending on whether your view of humanity tilts toward optimism or a misanthropic pessimism.)

But in fact there are gradations in vulnerability, and recognizing this continuity is crucial. It allows us to detect subtle individual differences that we would otherwise ignore. The distribution of susceptibility resembles a Gaussian curve, with most people falling in the middle and some people at the outer edges. Some people are extremely dogmatic and some are extremely undogmatic. Acknowledging that suggestibility is continuous reminds us that few people are completely immune to the power of ideological reasoning and community. If we lie on a spectrum, we must confront our own vulnerability.

Our place on the spectrum is not fixed. We can all shift our positions. In fact, we all do: we slide toward heightened ideological susceptibility in moments of stress and glide away from ideological solutions in times of comfort, of exploration, of discovery and adventure.

To visualize how a person drifts toward greater and greater ideological dogmatism, I often imagine transforming that spectrum—the continuous line that represents differences in susceptibility, from low to high—and bending it so that the right side of the spectrum curls inward and inward to make a sinister spiral.

Why a spiral? Spirals are magnificent natural shapes found in snail shells, sunflower seeds, artichoke leaves, a chameleon's tail, pineapple and pine cone scales, and tropical hurricanes. Spirals track development and movement over time.

In thinking about how a person shifts along the spectrum of ideological thinking over time, a spiral offers several interesting features. A spiral effectively captures the trajectory into an ideology, which may initially require slow, large increments on the outer surface (big leaps of faith) that are followed by a tighter and tighter coil, curling in faster and faster. We see this in the acceleration of cultish behavior—once a person passes the threshold of acceptance, the cult's demands for sacrifice become more exacting and more severe. A converted individual is suddenly eager to engage in actions that they recently found abominable. As the ideology renders the believer's brain more and more dogmatic, there is a self-reinforcing effect: every movement toward extremism becomes easier and smoother the deeper in you already are.

The spiral reflects the interaction between a person's dispositions and their ideological environment. In a given community, everyone will be positioned at different starting points, but the environment they choose (or are forced into) will affect how rapidly the person will adopt the most extreme conclusions of an ideology.

A highly flexible, emotionally regulated person may be on the spiral's outer edge. Relaxed, content with ambiguity, switching between modes of thinking easily, unlikely to commit to harsh dogmas. But provoked by a situation, or exposed to an alluring ideology, they may be propelled further in, moving toward greater intolerance, fixity, and hostility. Crucially, the effect of a trigger—stress, conflict, precarity, loss, the many terrible events and conditions that render a life less secure—will be starker for the person who is already highly susceptible, positioned close to the spiral's center point. For the person more suggestible at baseline, a triggering event or situation will push them toward the extremist core faster and more strongly. Ideological thinking will prey on their preexisting propensities. A rigid ideology will attract and satisfy the rigid, emotionally volatile mind more fully than it would for the flexible, emotionally secure mind.

By turning the line into a more mature two-dimensional spiral, we add time. We think about how a single person as well as entire societies evolve and react to their circumstances. We see how ideological thinking is a dynamic product of psychological traits and ideological experiences.

Mathematically, there are multiple types of spirals. There is the Archimedean spiral, where the distance between the curves is the same for each rotation. This is the spiral of a looped rope or a tight scroll compressed evenly, linearly, into nearly concentric circles. It is the type of spiral traced invisibly by a record player's delicate needle as the vinyl disc turns, the crescent grooves revealing the acoustics of a song. The curves are evenly spaced, the inward twist humble and space-efficient.

A second class of spirals is the logarithmic spiral, where the distance between curves decreases as the line curls inward toward its middle point. This is the spiral we see in snail shells that twist shyly inward or in the flow of water into a drain. It is also the shape of the dramatic rotations we would see if we sat in satellites looking at the earth from above to observe cloudy cyclones turning and speeding in the skies. If we rolled over and gazed beyond the earth's atmosphere at cosmic disc galaxies, we would see these logarithmic spiral arms once more. Their geometrical elegance enchanted René Descartes and prompted the famed seventeenth-century Swiss mathematician Jacob Bernoulli to call the curve *spira mirabilis*, miraculous spiral. In logarithmic spirals, there is an increasingly pronounced tightening as one follows the arc inward.

I imagine the spiral into ideological extremism to approximate this latter, logarithmic-style spiral, which accelerates as one is pulled in. Once sucked into ideological logic and community, it becomes easier and easier to get drawn more deeply inward—and more difficult to come out.

One's current position on the spectrum of susceptibility to ideological thinking is a constant interplay between one's vulnerabilities and the ideological doctrine. Baseline traits predispose a level of vulnerability, but every persuasive encounter can push the person further down the spectrum, into the spiral's eye. This is how tyrannical logics and tyrannical emotions work. They are never at rest.

Being immersed in the ideology changes the believer; it changes their explicit and implicit beliefs but also their broader cognition, instinctual responses, physiology, and brain as a whole. Predispositions and ideological communities reinforce each other, leading to the expression and accentuation of certain traits and the overriding of others. For instance, an inflexible and emotionally volatile individual who is enveloped by ideologies that prey on fears of disorder and upheaval may become especially dogmatic relative to the individual in a more moderate community. Risk factors become more pronounced when the ideological environment elicits and reinforces them. This is one of the ways ideologies beget the realities they prophesize: inflexible and catastrophizing doctrines sculpt adherents in their own image—crafting inflexible and catastrophizing minds.

This is also why de-spiraling is so difficult. Even if an ideological group dissolves and breaks apart, or if its contradictions and injuries become intolerable to the believer, the believer is already changed. They have become habituated to routines, rituals, a certain kind of reverence, surveillance, and insensitivity to their doubts. For the believer who has lost faith but not the logical and emotional dependencies of ideological thinking, it is therefore difficult not to slide down another spiral for a different dogma. As Eric Hoffer writes in *The Passionate State of Mind*, "The dislocation involved in switching from one passion to another—even its very opposite—is less clear than one would expect. There is a basic similarity in the make-up of all passionate minds. The sinner who turns saint undergoes no more drastic transformation than the lecher who turns miser."

The spiral into extremism helps us understand the different kinds of instabilities that are at play in ideological thinking. It depicts the instability that propels believers toward greater and greater dogmatism, allowing ideologies and cults to excuse increasingly violent actions and abuses within the group or against detractors. It discloses the instability that leads some lapsed believers—even if torn away from the ideological group to which they were committed—to search for another system to adopt passionately. The spiral also shows the instability of even the slight partisan: one push—a trigger or a crisis—and a whole sequence of cognitive and ideological pressures is enacted on the individual's body and brain.

Have I replaced one set of metaphors with another? Substituted the mindlessness metaphor of a passive believer with an image of a spiraling body sculpting itself, changing itself, being molded by the ideology it consumes? Maybe. Yet as George Eliot observed, "It is astonishing what a different result one gets by changing the metaphor!" This metaphor gives us a more mechanistic perspective and more explanatory depth. It does not shy away from layers, processes, interactions, and qualifications. This metaphor offers us richer opportunities for scientific investigation and falsification. It forces us to confront the origins and consequences of ideological thinking and bind the two together over time to model how the ideological brain evolves and changes. The spiral into extremism suggests why ideological thinking is not fixed or inevitable or stable or still. It suggests how a brain can become ideological but can also, with effort, crawl out of the ideological vortex it has been pulled into.

What accelerates or decelerates (or reverses) a person's trajectory into ideological extremity and intensity? What kinds of situations push a mind further in? Which situations allow us to climb out and sustainably stay out? What happens to vulnerable minds in vulnerable environments?

Our bodies are most vulnerable under the weight of stress. A stressed person naturally rigidifies, their attention narrows to the stressor, and their tolerance for uncertainty shrinks. Experiments that stress the human body in an acute but momentary way have often found that participants' cognitive flexibility suffers. One method to induce stress is the Trier Social Stress Test, in which the participant is given three minutes to prepare to deliver an oral presentation to a panel of grumpy interviewers and then complete an arithmetic pop quiz in front of the unenthusiastic panel. This sequence is truly a nightmare—unannounced public speaking, mental math, and social judgment by strangers—the body's stress reactions always spike. Researchers have shown that after completing the Trier Social Stress Test, people perform poorly on cognitive tasks that require adaptability, inventiveness, breaking habits, or switching between modes of thinking.

In other experiments, the stress induction is directly physical rather than social or psychological. How do you induce bodily stress in an ethical way? Plunge a participant's arm into a bucket of ice water! Participants in the stress condition—who must keep their arms in ice-cold water for what to them must feel like eternity (it lasts only three minutes) before completing a series of cognitive tasks—exhibit poorer flexibility than participants in the no-stress condition, whose arms float in buckets of lovely warm water before the flexibility test.

Interestingly, some researchers are finding that acute stress impairs cognitive flexibility primarily in men and not in women. This may be due to sex differences in the functioning of stress mechanisms, but the jury is still out. Since most studies on the effects of stress on neurocognition have been done in all-male participant samples, there are gaps and biases in how neuroscientists understand the female body's reactions to stress.

Babies feel it too. A small study with fifteen-month-old infants showed that stress impairs their cognitive flexibility. Half of the infants were exposed to a series of stressful situations—including separation from their caregivers and interactions with strange objects and unfamiliar people—and half of the infants were not exposed to the stressors, instead allowed to play with their caregivers in a calm setting. The researchers checked that the stressful situations were indeed physiologically stressful by measuring the infants' cortisol levels, which rose after the stress manipulation. The infants were then given a learning task in which they learned to perform a habitual action (pressing blue and red buttons in a sequence) that resulted in dazzling lights and musical tones. For a baby, that's a party. After a while, the flashing lights and sounds no longer appeared when the habit was enacted. Which babies would continue to press the buttons in the habitual sequence even though the pattern no longer led to the rewards? Once the habit was no longer effective, infants in the no-stress condition disengaged from the habituated buttons in favor of trying out other options. These fifteen-month-olds freely explored new possibilities. In contrast, the stressed infants stuck to the old no-longer-effective habits and refrained from exploring alternatives. Intriguingly, stress did not impair general learning or the babies' interest in the game, but it did damage the flexibility with which the infants were able to disengage from the old rule and engage with new possibilities.

New lines of research indicate that the effect of stress on decision-making is not the same for everyone. The decision-making of individuals with greater working memory capacity—a proxy for general cognitive ability and fluid intelligence—is less likely to suffer from an acute stressor. In contrast, when an individual with a smaller cognitive capacity is exposed to a physiological stressor (immersing their hand in the dreaded ice bucket), and their cortisol spikes, their flexible decision-making will be more severely compromised than that of their peers with better baseline cognitive ability. Stress skews people's decision-making to different degrees.

The next step for scientists will be to link all these pieces of evidence together to construct a fuller bridge between the neurophysiology of stress and ideology. How do stressful environments impact ideological thinking by way of altering neurocognitive processes? For now, research suggests that there is an interaction between the person's baseline cognitive traits and how they respond to the stressful stimulus. Minds prone to rigid thinking will experience worse decision outcomes under stress. A person positioned further along the spectrum of suggestibility will spiral toward dogmatism under stress, in a way that is faster and more pronounced than for a person who is more resilient to begin with.

Importantly, there is a difference between the effects of acute, brief stress and stress sustained over longer periods. A body that has been incessantly stressed over weeks and months and years will be more severely and permanently affected than a body that has been stressed for a moment and then returns to privileged tranquility. Notably, the longer a person is exposed to stress, the harder it becomes to measure it with precision or to induce it ethically in psychological experiments. One study used a natural experiment by comparing medical students undergoing a six-week stress-heavy period preparing for grueling exams with students experiencing six weeks that were relatively stress-free, studying medicine without exams. The researchers noticed that the decision-making of stressed participants became skewed toward more inflexible habitual behavioral patterns, and their brain morphology changed concomitantly too. Importantly, after the stressor was alleviated—the exams were done and dusted—the stress-induced changes to the structure and function of the brain were reversed. This suggests that the effects of stress on rigidity are plastic; the brain "bounces back" to its calm, flexible, explorative self.

If it is possible to observe the effects of stress on rigid thinking in anxious medical students, in infants briefly separated from their caregivers, or in shivering participants submerging their jittering fingers in ice-cold water, then the consequences of being immersed in environments in which there is sustained, unrelenting stress must be profound and troubling.

18

THE IMPORTANCE OF BEING NESTED

In tracing the spiral of ideological extremism, we are reinterpreting the problem of the chicken and the egg. Instead of probing what comes first and what comes second, we are asking how causes and consequences dynamically affect each other over time. Instead of getting stuck in an intractable circular conundrum, we are acknowledging that some neurocognitive predispositions push people into the arms of rigid ideologies, and that, at the same time, the passionate embrace of rigid ideologies can infuse the cognitive style and physiological sensitivities of the believer. Through this sinister feedback loop, individuals can become more and more dogmatic, more and more intolerant, their bodies molded by the ideologies they use to reason and feel.

Stress can accelerate these trajectories, rigidifying our thoughts and rendering us less adaptable, more belligerent, and more prone to habits. But stress is not only transient; it can be systemic. Some settings are marked by sustained pressures and restrictions while other settings are more serene and plentiful. To understand the interplay between the origins and outcomes of ideological thinking, we must attend to the environment in which these causes and consequences are taking shape.

In other words, a full solution to the chicken-and-egg problem requires turning to the *nest*. The ecology. The surrounding social context that envelops the individual as they develop their personality repertoire and ideological predilections. The nest is the home, the neighborhood, the city, the country, the climate, the pressures that are exerted on the person. The environment that places some people in situations of risk and others under the protection of resilience.

What's in a nest?

Nests are calming, protective cocoons meant to comfort. So what happens when individuals are embedded in stressful, as opposed to peaceful, environments?

Nests are also communal, designed to be shared. So what happens when individuals are born into a social environment marked by inequality, insecurity, and insufficient resources?

Some nests offer refuge: weakening the links between psychological vulnerabilities and ideological outcomes. Other nests are places not of rest but of battle and survival: accelerating the spiral into extremism, letting cognitive risks translate into radical decisions.

In adding the layer of the nest, we are examining how context moderates the relationship between cognition and ideology. To moderate means to alter the course of a trajectory or the strength of a relationship. Some contexts will expedite the trajectory into extremism—pushing psychologically vulnerable minds toward toxic doctrines—and some contexts will subdue the allure of ideological dogmas—rendering them less effective, less seductive. Perhaps there is also a kind of nest that de-spirals minds out of ideological intolerance; the kind of nest we should endeavor to build.

While *context* can be a vague and amorphous concept, when we define and ground it thoughtfully context can be the third variable that unlocks the rest of the puzzle. "The truth is rarely pure and never simple," exclaims a character in Oscar Wilde's comedic play *The Importance of Being Earnest*; "modern life would be very tedious if it were either." A consideration of context reminds us that scientific truths are almost always contingent, tentative, rarely simple, and never pure.

By delving into the importance of the nest, we can ask when an environment safeguards and when it imperils. What happens to the ideological brain when it is exposed to dangers and threats? And what happens if the nest itself is the menace to be feared?

Ideologies offer the promise of "home," a place where one is accepted and understood. Home is the space where we know the rules for how to behave, what to believe, and whom to trust. But how does the brain respond when it feels excluded? What does it do when it feels lacking in kin, lacking in reciprocity and belonging?

In an innovative study with men of Moroccan descent living in Spain who espoused militant jihadist views, researchers were curious to see what ideologues' brains looked like when they were made to feel socially excluded. To induce feelings of exclusion, participants played a virtual ball-throwing game called Cyberball. Throughout the game, a virtual ball is thrown between the participant and three other virtual players, who were endowed with Spanish names and photos corresponding to the participant's age. For the duration of the game, "Jose," "Javier," and "Dani" are the participant's peers as the ball is thrown and received between the players. Unbeknownst to the participant, the other players are preprogrammed simulated agents that are designed to throw the ball to the participant only a fraction of the time. Participants in the control condition received the ball a quarter of the time; they were equal players in the game. Participants in the social exclusion condition received the ball twice at the start of the game and then never again, condemned to watch the game unfold without their involvement.

Forced to bear witness to their own ostracism and omission, the excluded participants watched the cyberball make its bouncing arcs across the screen but never into their own hands. With every throw of the ball, they might have hoped to be included on the next turn—but, alas, they were repeatedly rejected, repeatedly disappointed.

The scientists then studied the participants' neural responses to sacred values for which they are willing to die—such as expelling US forces from Muslim lands—and nonsacred values, which are prized but negotiable—such as fighting for Islamic teachings in Spanish schools. Remarkably, the experience of social exclusion led nonsacred values to evoke the same neural signature as sacred values. Suddenly, shunned, the brain imbues every value with significance and holiness. When one is feeling jilted, even peripheral values become values that one is willing to fight and die for.

Regardless of the content of an ideology, social exclusion is one of the most powerful predictors of justification of terrorism and extreme political actions, both for left-wing and right-wing causes. Excluded individuals are more likely to endorse terrorist organizations fighting for pro-democracy causes and are more willing to damage property in the name of terrorist groups fighting for animal rights or environmental protection. Students who feel socially excluded are more receptive to joining gangs and activist organizations. Excluded individuals who are characterized by a heightened sensitivity to rejections indicate greater willingness to engage with extreme groups, including established political groups as well as novel or fictitious organizations that participants encounter for the first time during the experiment. Experiences of exclusion can prompt people to bend their ideological beliefs to fit more radical groups in order to feel a sense of belonging and can lead to greater certainty about the moral goodness of their group's beliefs. Excluded individuals become more willing to fight and die for their group, especially if they already possess strong baseline psychological needs to belong to a collective. Feeling rejected can foster a readiness for radical action, even for values previously considered nonsacred and for new values in the name of entirely new terrorist organizations.

Loneliness gears the brain toward finding groups to fill the void. Existing allegiances become more pronounced under situations of stress and scarcity. In a sophisticated set of experiments, American researchers found that under conditions of scarcity—situations in which participants felt that financial resources were limited—White American participants display racially discriminatory behavior against Black people. One neuroimaging study investigated whether scarcity conditions produce changes to neural processing that explain this behavioral discrimination. Participants saw a series of faces—some Black and some White—and made hypothetical decisions about how much of $10 they would allocate to each face. In the control condition, participants were told that there was a money pot of $10 and that a random computer model would decide how much money the participant would receive to allocate to others. In fact, the allocation was rigged. All participants in the control condition received the full $10 to distribute to others, and so they should have felt unbelievably lucky—flush with cash. In the scarcity condition, participants were given the impression that out of a possible money pot of $100, they only received $10 to allocate to others. So while the amount of resources that a participant in a control condition was "given" was the same as that of a participant in the scarcity condition, there was a difference in the framing of whether this amount was a lot or a little. When resources felt scarce (when the participant believed they received $10 out of a possible $100), there was a clear anti-Black, pro-White bias in participants' allocation decisions. This prejudice was absent when resources seemed abundant.

Scarcity also produced stark neurobiological changes. Chillingly, the researchers found that scarcity affected neural face perception mechanisms such that it took White participants longer to recognize Black faces as human faces. This was indexed by EEG signals that found a delayed response in the N170 waveform that typically corresponds to the detection of a human face. Moreover, an fMRI analysis revealed that, under conditions of scarcity, Black faces elicited less activity in the fusiform gyrus, the area of the brain typically activated by the perception of human faces. Notably, scarcity selectively impaired the neural encoding of Black faces and did not affect that of White faces. The behavioral and neural prejudices disappeared when financial resources were framed as plentiful.

"The deficits in face processing observed in this research, which were specific to Black faces viewed under conditions of scarcity, may represent a very literal form of dehumanization," reported the scientists. Ideologies about race, inequality, and social hierarchies are not merely political phenomena—they can seep into the most fundamental biological mechanisms in the brain, such as perceiving a face to be a human face. Discrimination and dehumanization are processes *of* the body. "The Other fixes me with his gaze, his gestures and attitude, the same way you fix a preparation with a dye," wrote the Martinican philosopher Frantz Fanon of the White gaze on the Black subject. "I am not given a second chance. I am overdetermined from the outside. I am a slave not to the 'idea' others have of me, but to my appearance . . . the white gaze, the only valid one, is already dissecting me. I am *fixed*."

Dehumanization becomes a cyclical, literal process. In situations of scarcity, minorities disproportionately suffer—and the brains of their judging witnesses contain this dehumanization process within. Political neuroscience that considers the role of context can therefore unmask processes that purely behavioral studies would miss. This science also gives us clues about the harrowing effectiveness of ideological rhetoric that frames issues around scarce economic and social resources. Scarcity can draw out the racist, the dogmatic, the fundamentalist, the fearful within us.

Stress can emerge from direct personal experiences of precarity, exclusion, abuse, or the threat of violence. But pressures also trickle into the body from larger macrocosmic experiences: wars, pandemics, natural disasters, risks to a community's physical safety or existential security. Threats to our existence can affect the spiral into extremism. The fear of death can push the body further toward intense ideological convictions.

One thinker who considered this issue in earnest was the cultural anthropologist Ernest Becker. "Of all things that move man," he famously wrote, "one of the principal ones is his terror of death." The fear of death drives human beings to deny death through "immortality formulas." Immortality formulas may be beliefs in an afterlife that extends our existence indefinitely. Immortality formulas may be "myths of heroic self-transcendence" in which our mortality melts into timeless glory, and we gather the courage to face death. *The brave soldier. The fearless poet.* Immortality formulas are developed by ideologies to attract and sustain believers. To justify self-sacrifice, ideological actions are imbued with grand and expansive significance. *O savior of the forests! O sacrificial freedom fighter!* Such labels of everlasting impact serve the objectives of ideologies well and allow adherents to disguise the littleness of life. Temporal lives are alchemically transformed into eternal essences. In its best guise, charity is an immortality formula, making us feel important beyond the confines of our lives. In its worst guise, violence or suicide in the name of an ideological cause is an immortality formula too.

Becker's theory was elaborated into what is now called "terror management theory": the idea that our worldviews function to control our terror of death. The human desire to deny death motivates us to seek and stick to worldviews that alleviate our anxieties about mortality and offer us a pathway to immortality instead. Through ideologies, terror management theory proposes, we manage our terror, we manage our trepidation of our life's end. As a theory, it is both powerful and a little absurd. "Management" evokes the formal, the organizational, the flow chart. Simply supervise and steer your terror! As though it is straightforward to stabilize the trembling hand or to steady the quaking heart.

Terror management theory has inspired a search for empirical evidence to confirm Becker's predictions. Hundreds of experiments have been devised to study the hypothesis that in times of existential stress, when reminded of the fragility of life and our impending, always-premature death, we will grip ideologies tightly for support. There have been two competing variants of this hypothesis. One variant, called the *worldview-defense* hypothesis, speculates that when faced with our finitude, we will become more defensive of our cherished worldviews. Reminders of death will render us more ideologically extreme, regardless of the mission of the ideology, regardless of whether we lean politically left or right. The other conjecture suggests that reminders of death prompt a shift toward conservatism. After all, a threat should trigger us to conserve, to preserve, to hold on to knowns rather than venture into more liberal unknowns. This is the *conservative-shift* hypothesis.

To provoke a sense of existential dread, scientists have designed clever and at times morbid manipulations to enhance the salience of mortality. These experiments typically invite the participant to dwell on their death—*really dwell on it!* Write in detail about what you expect will happen to you when you physically die. Write at length about the emotions that thinking about death arouses in you. Imagine the unimaginable end. Some investigators study naturally occurring moments when death is prominently on people's minds: national catastrophes, violent conflicts, raging epidemics, or the traumatic aftermath of terrorist attacks. Other experiments take place in or near cemeteries, and some scientists have even concocted virtual-reality experiments in which participants take a simulated walk through a graveyard.

It is a creepy experience. Maybe better not to dwell on it! But for those who did dwell on their death, heightened mortality salience sparked greater attachment to their chosen ideology. Compared to participants who were asked to write about neutral topics or imagine other types of pain unrelated to death, such as a throbbing toothache (though who doesn't contemplate death and violence while in a dentist's chair?), the participants who were instructed to write about their death were the most ideological after the manipulation.

Across these different procedures and methods, meta-analyses combining the results of hundreds of studies reveal that mortality salience prompts worldview defense. Iranian college students randomly assigned to the mortality salience condition positively evaluated a student who justified martyrdom whereas those assigned to describing dental pain were suspicious of the martyr and preferred the responses of an anti-martyr who displayed tolerance and aversion to violence. Conservative American college students who were randomly assigned to the mortality salience manipulation supported more extreme military interventions, such as preemptive attacks on foreign countries, killing thousands of innocent civilians, forsaking personal freedoms to enhance national security and military might. For environmentalists, mortality salience heightened concern for the environment, and emphasizing climate threats led German environmentalists to be more authoritarian and conformity-minded regarding other group members. Reminders of death can propel people to believe in more extreme, violent, and retaliatory ideological solutions that affirm their prior beliefs. But meta-analyses demonstrated that mortality salience also induces conservative shifts to the right, whereby more traditional authoritarian leaders and policies were favored by both political conservatives and liberals. How can both the worldview-defense and conservative-shift hypotheses be true?

Perhaps in the same way that we see the rigidity-of-the-extremes and the rigidity-of-the-right effects coexist in the link between cognitive rigidity and ideological thinking, here we see that existential anxiety has an affinity with right-wing orientations but can also elicit strong defense of one's existing ideology, regardless of what it is.

Predicting when a triggering threat will push people toward general conservatism or toward extremism (or when it will have no effect on ideological allegiances) may depend on the specific nature of the threat—whether it is directly existential or atmospheric, whether it is a politicized threat or a novel one. Terrorist threats often lead to shifts toward conservative and militarized solutions and increased hostility against outgroups, especially Muslim religious outgroups in cases of Islamist terrorist attacks. The reaction can also depend on the charisma of relevant leaders; following a mortality salience manipulation, people gravitate toward leaders who exhibit steadfast optimism rather than more humble and modest aspirations. After a threat, people love visionaries and miracles.

Some critics of the mortality salience manipulation wonder whether it works consistently or if it works for all people. While there have been some failures to replicate the mortality salience effects, there is also the possibility that the effects are not homogeneous or universal and require a more detailed and personalized analysis. For instance, there is evidence that the manipulations work especially well for people with low tolerance for ambiguity and strong needs for structure and meaning in their lives. Vulnerable minds are the most affected by rigidifying contexts.

There are also instances when reflecting on death may not produce existential anxiety but rather generate existential awe. In cases of near-death experiences or diagnoses of terminal illness, instead of terror, a person may experience a sense of transcendence. In the face of imminent or imagined death, a person may feel not only apprehension but also choose to reprioritize their values away from resource-hoarding greed and toward altruism, generosity, and building closer connections with others. Some cultures are better versed than others at channeling existential fears toward harmonious and peaceful worldviews. And this is the challenge that future politics faces: how to translate threat into creativity rather than fear or conformity.

Ernest Becker's *The Denial of Death* is peppered with the language of quests, of heroism, of overcoming the fear of death. But perhaps an attraction to ideologies stems not only from a bleak quest against death but also through a positive quest for a sense of personal mortal significance. One research program has shown that vulnerability to engaging in extremism emerges from a search for meaning. The investigators studied settings of adversity and hotbeds for radicalization. In Islamist militants belonging to the Filipino jihadist group Abu Sayyaf, diminished personal significance was linked to greater religious fundamentalism. In militants in Sri Lanka and the Philippines, feelings of personal failure contributed to more extremist and violent convictions. These patterns have been replicated in American contexts too. When looking at criminals convicted of ideologically driven crimes in the US, the ones most likely to have resorted to violence for far-right, far-left, or Islamist ideologies tended to have experienced rejection or failure at work, in friendships, or in romantic relationships. Inducing feelings of significance loss by asking US college students to reflect on moments of humiliation led them to endorse their worldview more emphatically: Republican students adhered more strongly to Republican values and Democratic students identified more passionately with Democratic values. Probing the reverse process of significance *gain*, researchers found that engaging in more radical political acts produced a greater sense of personal significance for activists than participating in milder and more moderate forms of engagement. This was true for environmental activists as well as labor and feminist activists. The quest for significance drives people to commit to more radical, demanding, and sometimes violent behaviors.

As described by the spiral into extremism, the link between these psychological vulnerabilities and ideological radicalism becomes exacerbated in certain contexts and subdued in others. Differences between cities and even neighborhoods lead to substantial differences in risk. In a study of over 4,000 school students in the Andalusian region of Spain, adolescents' support for violent narratives was predicted by the degree to which they perceived a lack of personal significance and felt undervalued or lonely. Endorsement of aggressive behavior and the belief that violence is necessary for social change was also explained by the degree to which the students belonged to a network of deviant peers who glorified fighting. Beyond these psychological and social factors, the broader economic and cultural environment mattered. The most vulnerable students lived in towns or neighborhoods marked by lower parental education and household incomes, fewer books available at home, and a scarcity of resources and opportunities. These children were the most amenable to the toxic effect of deviant peers. In neighborhoods with more prosperous economic and cultural resources, the link between befriending aggressive friends and possessing violent beliefs was weaker.

A well-resourced environment buffers against risky psychological states and friendship groups. This pattern is supported by other research efforts showing that Muslim high school students living in vulnerable environments in Spain, in towns known for being hubs for terrorist recruitment, were more likely to feel socially excluded, to perceive greater conflict in the relationships between Muslim and Christian students, and to legitimize terrorism than Muslim students living in less vulnerable towns. Similar studies with adolescents in Sweden and adults in Indonesia, Sri Lanka, and Morocco found that living in vulnerable social contexts strengthened the link between pursuing significance and supporting violence.

Vulnerable environments therefore do not merely add risk, they also lead to the accentuation and amplification of other risks. In thinking about the spiral into extremism, we can consider how the dynamics between neurocognitive vulnerabilities and dogmatic ideologies can become stronger and faster in settings of heightened stress and precarity. Neither the chick nor the egg can be understood without a sense of the nest.

Digital environments can also be sites of enhanced vulnerability. If the information circulating on these platforms is skewed, selective, or sparse—primarily presenting confirmatory evidence and excluding inconvenient information that we would rather scroll past—even the rational and undogmatic citizen will develop biased beliefs. Brains update their beliefs in response to the information that circulates in their environment. When information is unreliable or easily faked, it can lead to downstream distortions that are psychologically rational, even if they diverge from reality. As technological tools become ever more sophisticated in creating false pictures of reality, it will become easier for malicious agents to warp the rational and emotional apparatus of even the most intelligent thinkers.

Although new kinds of cognitive distortions may arise on digital platforms, the phenomena are not historically novel. The Polish dissident and poet Czesław Miłosz wrote in his 1953 book *The Captive Mind* of the person embedded in the propagandist echo chambers of the Soviet Union:

> A man need not be a Stalinist to reason thus . . . The propaganda to which he is subjected tries by every means to prove that Nazism and Americanism are identical in that they are products of the same economic conditions. He believes this propaganda only slightly less than the average American believes the journalists who assure him that Hitlerism and Stalinism are one and the same. Even if he stands on a higher rung of the hierarchy and so has access to information about

> the West, he is still unable to weigh the relative strength and weakness of that half of the world. The optical instrument he sees through is so constructed that it encompasses only predetermined fields of vision. Looking through it, he beholds only what he expected to see.

The principles of tightly controlled propagandist states are not unlike the ones experienced in contemporary digital life. Online or offline, is it a difference in degree or in kind? There are some unique features of virtual environments. Many digital platform algorithms are now engineered to spread the most emotionally dysregulating information, personalized to confirm beliefs and anxieties. Take a vulnerable mind—sensitive to negative information and threats, cognitively rigid, impulsive—and place this mind in an environment that selectively preys on these biases, and it becomes possible to understand why these spaces, though toxic for everyone, are especially toxic for people who are at baseline already psychologically vulnerable. The more technology obfuscates the source, truth, and reliability of images, texts, videos, and reports, the more likely it is that digital environments will be radicalizing spaces.

Just as vulnerability can be heightened in certain places, it can also be amplified during particular time periods. Adolescence is a distinctively vulnerable time—an age of "fleeing the nest" (or at least trying very hard to flee). It is a sensitive time for brain development and the evaluation of risk, conformity, flexibility, and the crystallization of identity. Ideologies often work on and through the young. Every nationalist movement has a youth chapter. Every religion begins its indoctrination in childhood (and forces converting adults to go through the religious child's ceremonial stages—conversion as rebirth). Almost every political movement that calls for a dismantling, a revolution that upends everything, recruits its earliest members from the ranks of adolescents, or adults prone to impulsive crises. The allure of alternative histories or alternative futures is often strongest for those who are negotiating meaning—whose brains are sorting through different predictive schemes with which to understand and anticipate the world and themselves.

How should we treat the young person who adopts rigid doctrines and rigid identities, who repeats clichés, obsesses passionately, nurtures dogmatic worldviews resistant to evidence, pursues the thrill and euphoria of togetherness even at the expense of freedom, longing to feel significant and loved? (In many ways, this description of radicalism is hard to distinguish from a general description of youth.)

A parent, a sibling, a friend may suddenly find a beloved child, sister, or peer committed to an ideological cause in ways that are hostile, intolerant, pugnacious, and morally arrogant. Whether or not we agree with the mission they are now fighting for, we may feel that our beloved has embraced a role that does not represent their authentic self. It is hard to find the line between caring concern and respecting freedom of choice. It is often unclear what is permanent and what will pass or fade—what our duty is to our dearest, what our duty is to society.

We can be compassionate and attentive to all the factors that lead a person to radicalism. But compassion does not need to turn to complicity. While it is true that the adolescent brain is the most susceptible to ideological dogmas, this is partly because it is a brain that is hyperactively seeking to understand the world and to be understood back. Adolescent brains thirst after a model of reality they can predict and participate in.

The demand for an explanatory framework is present in every adolescent. Part of the problem is what communities supply to satisfy this demand.

The political question is how we can design societies in which ideological solutions are not the only—or, at least, not the most salient—solutions for the brain's needs. Can we harness forms of political engagement and activism that do not inevitably fall into black-and-white ideologies? Can we offer citizens the tools to know when an ideology merits critique and when a philosophy can be a source of creativity and resistance to injustices?

19

OTHERWISE

Flexibility is a fragile thing. Even in the freest of places, ideological ways of thinking are terrifyingly alluring. Our predictive brains seek out rules, logics, and habits that can organize ourselves and others. It is a constant struggle to shun black-and-white thinking in favor of seeing all the shades of gray. Remaining in the liminal spaces of ambiguity is an arduous Sisyphean task—always demanding our attention, rarely allowing moments of rest.

In conditions of oppression, tyranny, or precarity, these demands can be even higher. If ideological restrictions are strong and powerful, it can be exceptionally difficult to resist, mentally or publicly.

Yet freedom outside does not guarantee flexibility within. "There is a large measure of totalitarianism even in the freest of free societies," wrote Eric Hoffer, "but in a free society totalitarianism is not imposed from without but is implanted within the individual. There is a totalitarian regime inside every one of us."

It is therefore a mistake to think that democratic rights or secular societies alone can protect us from the dangers of dogmatism. Although political emancipation and plurality are fundamental to flexible thinking, they are not sufficient.

In many respects, the freer our societies, the more important our individual traits and choices become. When we live in environments where many alternative ideologies are on offer, where many different ways of living are possible, then our psychological tendencies become potent predictors of our ideological behavior. Our personalities, cognitive dispositions, and physiological reactivities matter greatly in such settings, influencing whether we endorse philosophies that encourage people to think malleably or ideologies that discourage critical thought.

A study of individuals in free societies reveals that there are psychological and biological traits that render some of us more susceptible to ideological systems and others less easily enticed. In developing an image of the ideological body, layers of tissue and levels of analysis have been traversed to uncover how the believer's nervous system affects the ideological habits they adopt and how ideological habits may skew a believer's body. At times, the ideologue's perceptual and physiological reactions become heightened or numbed. At times, the ideologue's love for rules spills outside the realm of morality or politics and leaks into every private perception and rigid interpretation. The ideological brain is a brain that is cognitively rigid, emotionally dysregulated, physiologically less sensitive to injustice and injury, neurobiologically receptive to addictive rituals and binary categories.

But not all ideological brains are identical. Each person is a cocktail of particular combinations of personality and biological attributes. Some people will have certain clusters of risk traits and lack others. A brain that is amenable to adopting rigid doctrines and rigid identities may not, in the end, act upon and express these vulnerabilities.

Protective characteristics, such as cognitive flexibility or emotional regulation or a resilient familial context, may suppress or weaken the effects of risk factors. Else Frenkel-Brunswik witnessed such interactions between factors in her sample of prejudiced and unprejudiced children who completed tests of psychological rigidity. She remarked on "some exceptional cases of children who are extremely high on the prejudice scale but manifest only an average amount of mental rigidity." For children who were not strikingly rigid, it is possible that prejudice may have stemmed from other traits, experiences, or environmental factors, such as being embedded in an intolerant household or enduring other kinds of exclusions and challenges to their self-esteem.

There were also instances of the opposite pattern: when a child's psychological rigidity is higher than one would expect from their low levels of overt prejudice. Frenkel-Brunswik described "a girl who was most articulate in proclaiming a liberal ideology and who at the same time displayed more mental rigidity than is common in the unprejudiced." After interviewing the child's parents, Frenkel-Brunswik found out that the girl belonged to "a family which, though clinging in a dogmatic and militant way to a liberal ideology, did so with a great deal of inexorability and a lack of willingness to arbitrate with, or to accept, those who thought differently."

How we think can matter more than *what* we think, and by uncoupling the substance and structure of ideological thinking we can figure out when the content of the beliefs matters too.

While there seem to be cognitive and personality characteristics that render a person susceptible to *any* extremism, there are also traits that may spur an individual toward a particular kind of ideology. Some minds may be particularly attracted to ideologies that champion hierarchies while others prefer egalitarian structures. Some brains have an affinity to ideologies that taste of revolution and others desire nothing that evokes disruption. It may be that a brain with impulsive tendencies will specifically gravitate to ideologies that resist majority opinions rather than ideologies that affirm the status quo. The ideological brain attuned to religious and spiritual worldviews has a particular neurological architecture that is different from the ideological brain that disputes all supernatural explanations. There are unique factors that predict attraction to particular ideological doctrines as well as universal factors that predict attraction to any dogmatic system of beliefs.

In many ways, the science of the ideological brain raises as many questions about the nature of the brain as it does about ideologies. If there are parallels between an individual's perceptual and physiological apparatus and their prejudiced tendencies, what causes these parallels? Why do these echoes of rigidity exist in such different domains and at such different timescales? What does this tell us about the unities of perception and personality? How do rigidities permeate multiple levels of consciousness—in fast perceptual decisions, linguistic imagination, and political evaluation? To the extent that consciousness emerges from nested hierarchies of information processing and learning, future research will need to address how cognitive styles become reflected in these hierarchies. We need biological models of consciousness that can explain these structural resonances between unconscious split-second patterns and conscious convictions.

These linkages between perceptual flexibility and political flexibility, between creativity and intellectual humility, may also mean that when we exercise one type of flexibility, we may be fostering other kinds of flexibility at the same time. To teach a child to think malleably in one scenario can help them to think malleably in other scenarios too. The novelist Zadie Smith underscored this when she wrote that "flexibility of voice leads to a flexibility in all things."

Sometimes it is assumed that engaging in a classically creative pursuit—such as art, performance, craft, music, or literature—is equivalent to engaging with it creatively. But a creativity that matters psychologically is a creativity that changes forms and destabilizes, that breaks or stretches the convention, that experiments with inversions and reversals. This is our personal capacity to think otherwise—"*otherwise* as in, a firm embrace of the unknowable," in the words of writer Lola Olufemi. Embodying this flexibility is not necessarily constrained to the traditional arts; it is possible to rehearse a multiperspectival and adaptive creativity in every domain. The key is to switch between modes of thought and expression rather than rehash formulas. Even the person enclosed by repression and constraint can break routines and bend them in novel and liberating directions.

Many ideologies purport to promote creativity within them. Militarized groups parade and march to perfectly executed music. Every social movement asks protesters to produce inventive slogans and eye-catching decorations. Patriarchy convinces women to become ornaments in ways that ultimately disable them. Ideologies frequently play with art and aesthetics, but there are always limits on how far the believer's imagination is allowed to stray. It is only outside of an ideological framework that individuals possess the freedom to imagine and reimagine themselves over and over again. As the writer and civil rights advocate James Baldwin observed, "the artist cannot and must not take anything for granted, but must drive to the heart of every answer and expose the question the answer hides." This is why ideological systems fear (and sometimes incarcerate) transgressive artists, poets, and critical thinkers—they question the ideology's utopian answers and erode the doctrine's monopoly on truth.

For some people, rigidity is not a flaw but a virtue. Persistence and perseverance are perceived as essential for achieving our goals. The self-imposition of habits and routines is marketed as the path to success and well-being. After all, how could we—as individuals and societies—accomplish anything if we didn't stick to our guns? But a perspective that reveres rigidity typically disregards the nuances of stubbornness. In cognitive tests of rigidity, there is no advantage to inflexible behavior—in fact, it is a disadvantage. Inflexibility exemplifies a lack of adaptability, an absence of inventiveness, an insensitivity to changing evidence. These rigidities translate into dogmatisms that bruise our minds and bodies.

But there are ideologies of freedom, I hear the critic say, *ideologies of liberation, ideologies of love. These are not the same as ideologies of tradition and exploitation and closedness.*

I think that an ideology of freedom is an oxymoron, a self-contradiction. There can be a *philosophy* of freedom—a vision for what autonomy looks like—but the moment it becomes a systematic doctrine with an irrefutable pseudoscientific and essentialist logic, it ceases to be about freedom.

The new science of the ideological brain should vitalize any philosophy that tries to define itself as malleable and oppositional to dogma. By reckoning with how easily a philosophy (or even a science) can become an ideology, we can inquire into how philosophies of freedom can avoid slipping into ideological systems.

Ideological thinking is allergic to a scientific outlook that is inquisitive and provisional, an outlook that is genuinely interested in conversation and revision. To be free and to break from an ideology is to engage with multiple voices, to play, and to reject the overly serious. To dare to go off-script.

I want to invite you to participate in one last experiment—the ultimate game, the final test.

Imagine you are sitting in front of a screen. Instructions will pop up shortly. As per ethics protocols and regulations, the instructions will remind you that you are free to leave the experiment at any time. You are allowed to express dissent or question the instructions. You may desert the role you have been given.

In fact, perhaps you should. It would be inconvenient for me as the scientist if you left the experiment now, as I want to measure and quantify you, to give you arbitrary rules and watch you try to stick to them.

But maybe you should get up. Step out of the room and walk freely. Know that you can leave. Remember that you can resist irrational rules handed to you.

I hope that you will drop rigidities on your way out, that you take the time, that you feel no pressure, no predestination, no ancestors on your shoulders or rituals to obey, no expectations weighing you down or obstructing your movement.

I hope that you will interrogate all "shoulds," all duties, all compulsions imposed upon you from outside, all prescriptions to behave in certain ways that leave you uncomfortable.

I hope that if you bow down for a moment, it will be to pick yourself up and walk away.

I hope that you reject rules and instructions, resist illegitimate authority, that you find a private, authentic kind of liberty and stride onward.

The choice is yours.

EPILOGUE

GOING OFF-SCRIPT

I put the script down and look up. The auditorium lights fade in like sunrise and the faces of the audience come into view. Hands shoot in the air—some confident, others timid—and the microphone is handed to a listener in the back of the hall.

Can the science of ideology help us to learn which ways of living are good and which are bad? the voice asks.

Yes, I speak in the direction of my interlocuter's silhouette. I believe there is a sense in which this science provokes us to reassess our preferences, our habits, the criteria by which we judge whether a worldview is oppressive or liberating. If our ideologies are interconnected with our biological realities, then the stakes of our ideological commitments and choices are much greater than we previously thought. The time-tested notion that "the personal is political" takes on another dimension of meaning. By studying the interplay between ideologies and human brains, we can inaugurate a new method for critique.

If I may follow up, the voice speaks again, *what do you mean by critique? Is that merely a fancy word for criticism or judgment?*

Critique is about judgment, yes, I acknowledge. Critique is a synthesis of two voices. The negative "critical" voice we attribute to the fault-finder, the killjoy. This voice is combined with the voice of the positive "critical" thinker who reasons independently of authority, who overcomes one-dimensionality and interrogates surface appearances and the dominant perspective. Critique joins the suspicion of the negative critic and the creativity of the positive critic to construct a full-bodied analysis of an ideology. Crucially, ideology critique has a view toward emancipation. It hopes to free others through its careful unpeeling of facades and its suggestions of alternative futures or interpretations.

But scientists don't do ideology critique, an accusatory scholar of the humanities resounds skeptically. *This is the task of philosophers, historians, economists, sociologists, and cultural thinkers.*

Why not? I ask, recalling that every aggression is an act of defense.

Because, an ancient baritone voice rises from the back row, accompanied by Darwin's grand silhouette, *it is not the scientist's place to critique ideology*, he admonishes. *Science and politics must stay separate. Scientists ought to keep their controversial ideas to themselves and their close relations, lest they offend and diminish the impact of their discoveries!*

There is truth in your words—I speak to Darwin's bearded profile—but by staying silent on certain topics, you have allowed some of your science to be twisted to justify racist, religious, and conservative ends, which I suspect you might privately abhor. There is a difference between a science with political and existential implications and a science designed to appease particular political convictions. A science can critique dogmatic practices without losing its objectivity, rigor, or falsifiability.

Oui, she's right. Someone else pipes in, an even older voice, French-accented and dressed in thick eighteenth-century robes. *Idéologie*, Count Destutt de Tracy says spiritedly, *was destined to be the science of our sensory and rational ideas. But it was always an educative project too, to channel the science of sensation and reason into enlightenment. I dreamed it as a means to unchain the French population from orthodoxies that should be long retired.*

Out of the corner of my eye, I notice Napoleon leaving the auditorium with a click of his heels. Before the door slams he cries toward the stage: *You are the enemy of the people, a thief of happiness. I cannot stand this any longer!*

As one diva leaves, another raises a clenched fist. Is it a question? An exclamation? A symbol of solidarity or defiance? Before I can point the roving microphone in its direction, the owner of the fist begins to speak.

I wrote about the ideological phantoms in the human brain long before these strange neuroimaging technologies were invented, Karl Marx announces. *It is no surprise, of course, but I'm glad to discover these phantoms are real.*

Did I ventriloquize Marx's self-congratulatory comment? Almost certainly.

But I think you are blind on one front, Marx sighs noisily. *Have you not thought of your positionality? Who are you to decree which ideology is good and which is problematic?*

This is why we need scientific approaches to inform ideology critique. So far, our methods for determining which ideologies are moral and which are immoral have relied on roughly five kinds of sources. First, historical analyses of past sufferings. Second, philosophical or theological pronouncements on universal abstract moral categories and whether these are being frustrated or fulfilled. Third, cultural comparisons of existing ideologies. The fourth method uses aggregate social and economic reports to study the population and test who fares better or worse and why. And finally, personal testimonies of the lived experiences of victims of unjust ideologies illuminate which opportunities a prevailing ideology limits and which it forecloses entirely.

These are excellent methods for critique! Marx stresses.

Indeed, they are. But this new science can add a sixth, unique approach to critique. It is designed not to eclipse the other methods but rather to complement them with another angle, another level of analysis: the ideologized individual, the brain and body that reflect the ideologies that enter them.

And what does your "new method," Marx pokes further, *genuinely show?*

That we can bypass self-reports of ideological oppression and that, instead of relying solely on the tools of history, economics, demographics, and ethnography, we can study unconscious processes that are inconspicuous to the observer's naked eye or the believer's storytelling tongue. After all, *you* should know that ideological oppressions are not always experienced as such. The most effective ideologies embolden followers to desire their own domination. With the tools of science we have a fresh way to distinguish ideologies that distort our cognitive capacities from worldviews that bring us closer to our sensation of reality. We can quantify the harm, the injury inflicted by rigid ideologies in new ways. This allows us to be attuned to changing currents and new ideological movements. For instance, by demonstrating that nationalism invites a kind of mental rigidity that spills into many corners of our cognition, we can question what kinds of national and patriotic groupings are healthy for the brain. Can we imagine group borders that elicit love, altruism, and cultural investment, without the erection of inflexible psychological borders and prejudiced attitudes? Is solidarity without ideology possible?

I am afraid I must protest, a muscular voice inflects in a more modern German.

Yes, Dr. Arendt? I search for her standing shadow outlined by the dusty hall backlights. But then I discover her closer than anticipated, on the edge of the front row, sitting, not standing, a plume of smoke emanating from a cigarette held by her cheek.

This hypothesis sounds very plausible and I think it is quite wrong, Hannah Arendt exclaims and lifts the cigarette to kiss her jagged teeth. A puff, and she continues. *I do not see that biology has anything to do with the political questions of totalitarianism and evil. Evil possesses neither depth nor any demonic dimension.*

Her lips are drawn into an elongated D shape, which I initially interpret as a grin but gradually I recognize as expressing not a smile but rather an impatient call to attention.

My question to you, she continues, *is how a cognitive and biological perspective helps us to judge radical or extreme evil.*

The question of evil is a misguided question, I reply. For me, the burning question is not whether an abstract evil exists. The issue is not whether the human heart is malevolent or kind or stupidly selfish. The question is one of process, and therefore a question of depth and consequence. The deeper into the layers of the ideologue's body we enter, the more we see. We find that ideological thinking is not shallow or surface-level. On the contrary. Ideologies twist thought patterns in systematic ways, across levels of consciousness, in the emotional cores of our brains as well as our more reflective organs. It is overly simplistic to attribute evil to an on–off switch of thinking and nonthinking, as you seem to imply: evil does not start and end when we cease or begin to think. It may be easy to say that what happens to the brainwashed individual is a nothing—a mindlessness, an emptiness. Yet the ideologies that cajole us into action, that push us to love and to hate, do not produce minds that are hollow and mysterious. Minds we cannot describe or fathom. Minds characterized by ignorance or a weakness of will. On the contrary. The brains of passionate believers are knowing brains—active and conscious and rational. Ideological brains are not impenetrable black boxes; they are knowable organs. This new science unveils the neural processes that produce dogmatic people, people bent on discipline and binaries.

But what does this mean for moral judgment and action? Arendt insists, switching between German and English. *You have not said how we develop a moral conscience, how we discover moral categories, how we monitor a moral compass. Does this science tell us anything about how and who we condemn? In fact, let me ask the question in your terminology,* she persists. *Is a brain that is susceptible to ideological thinking responsible for its extreme and hateful actions?*

In my view, responsibility does not evaporate, I answer, just because certain individuals are more vulnerable to dogmatic thinking than others. Risk factors are probabilities and potentials that are constraints on actions but not certainties. In every pool of participants who possess vulnerabilities, there is a sizable proportion who defy what a determinist might conclude is their destiny. If ideologies are learned and chosen by active brains, then the principles of learning and choice also govern their brains when they actively decide to rebel and revolt. The brain is malleable and plastic—yes, within limits, but it is plastic all the same—and society has a responsibility, in my opinion, to nurture this plasticity and treat its citizens as free agents.

An arm waves calmly from the center and I turn away from Arendt, whose furrowed expression remains an impenetrable mix of discontent and deliberation. The figure stands, and I immediately recognize the eyes from sepia-toned photographs—eyes that are indisputably wise but also streaked with defiant mischief.

In my own interviews and experiments in the 1940s, Else Frenkel-Brunswik begins, *one of the predictors of the authoritarian child was their agreement with apocalyptic statements such as the notion that some day a flood or earthquake will destroy everybody in the whole world. The children prone to rigidity were fascinated by chaos, upheaval, and catastrophe. In the desire for order there was a fetishization of disorder. Looking at your own time, these sentiments seem present again in seemingly opposing ideologies: religious cults, right-wing authoritarianism, but also left-wing environmental and anti-capitalist movements.*

I wonder whether you think that the authoritarian personality—maybe you would call it the authoritarian brain—is more prevalent in your era or mine. And what is to be done?

I begin to formulate my answer when I hear a child's nervous breathing sounding across the auditorium's speakers. A young person whom I do not recognize stands up at the back and clutches the microphone, grasping it almost desperately, maybe for safety, for comfort, for fear. Their voice trembles at the start but steadies with every word.

It sounds like you are advocating for the cancellation of all ideological beliefs, they say, *and with that the elimination of all activism, solidarity, coalition building, expression of identity interests, and communal resistance. If we have no ideological beliefs, how can we fight for change or progress? How can we cultivate moral clarity? Without ideologies, how can we imagine a better future?*

The young person's lips tighten. Eyes narrow. *Is it fair to conclude that you are calling for complacency and moderation?*

Not at all, I reply. I do not believe that the way to battle rigidity is to slide toward the center, to achieve a diluted, shrinking moderation. The center will always change. The center will never hold.

I think we need to consider what it means to develop ways of living and thinking—privately and together—that are nonideological. Ways of existing that resist rigid doctrines and rigid identities at every turn. I believe that defying rigidities requires us to envision what an *anti-ideological brain* might look like. An existence that actively and creatively rejects the temptations of dogma. A mind that is ideology-free.

ACKNOWLEDGMENTS

The freedom to think and create is the most profound privilege, and I am deeply grateful to all the people who have encouraged and facilitated this freedom.

Thank you to my phenomenal literary agent, Rebecca Carter, for your wisdom, sharpness, care, and flexibility in all things.

Thanks to the amazing Margaret Halton and the fantastic team at PEW Literary—Patrick Walsh, Alex Chernova, Terry Wong, Cora MacGregor, and Rebecca Sandell—for helping this book reach an international audience.

Thank you to my wonderful editors, Connor Brown and Tim Duggan, for your trust and championing of this work. The teams at Viking and Henry Holt and Company have shown the book such caring attention. Thank you to Zoë Affron and Olivia Mead for your support and energy throughout the journey. Thanks to Emma Brown, Richard Bravery, Christopher Sergio, Nicolette Seeback Ruggiero, my diligent copyeditor Trevor Horwood, and everyone who has worked behind the scenes to bring this book into the world.

Thank you to all the research institutes, funding agencies, and organizations that have given me the latitude to engage in this radical science. I feel huge gratitude to the University of Cambridge and the Department of Psychology as well as Downing College, Churchill College, and the Gates Cambridge Scholarship for supporting my development as a thinker and researcher. The Paris Institute for Advanced Study—located in one of the most beautiful places on this lovely Earth—has been a haven for thinking innovatively about the brain as it navigates and absorbs social norms. I am grateful for its support of my research. The Archive for the History of Sociology in Austria at the University of Graz has been immensely helpful for accessing archival material on Else Frenkel-Brunswik. Thanks to the David Herzog Fund for supporting this project. Moreover, I'd like to acknowledge the Institute for Advanced Study in Berlin for giving me the space, time, and freedom to begin writing this book. The Wissenschaftskolleg's advocacy for creative interdisciplinary thinking is exceptional and Berlin was the ideal place to start.

I'm grateful to my academic supervisors, mentors, colleagues, and peers who have supported this research directly and indirectly. Thank you to my doctoral supervisor, Trevor Robbins, and adviser, Jason Rentfrow, for the warm early support when these ideas were in their most nascent formations. Thanks to Russell Poldrack for supporting my time at Stanford and encouraging me to generate the most interesting and rigorous analysis. Thanks to Mina Cikara, Fiery Cushman, and Joshua Greene for the enriching conversations at Harvard. Thanks to Corinna Hasbach, Patrick Haggard, Manos Tsakiris, Ryan McKay, Jon Simons, Amit Goldenberg, Nigel Warburton, and Barry Everitt. Thanks to Steven Pinker for the early support of this book.

The Ideological Brain is a kind of love letter to the field of political psychology and neuroscience—a field that is new, exciting, and ever changing. I would like to thank all the researchers whose experiments and theories are explored throughout the book as well as the next generation of scholars who wish to ask fresh, bold questions about the intersections of the mind, body, and politics.

Thank you to my friends and family who have shown an infectious kind of enthusiasm, curiosity, and love as I wrote this book.

Freedom can rarely be enjoyed without love (the reverse might also be true), and I feel unbelievably lucky to experience both. To my brother and sister, thank you for bringing such joy, laughter, and brilliance to my life. To my dad, thank you for being the most flexible person I know. To my mom, my cherished intellectual companion in everything, every good idea is dedicated to you. Thank you for creating a life so rich in possibilities for experimentation and happiness. To my incredible partner, thank you for creating and sharing this beautiful life with me.

NOTES

EPIGRAPH

ix **All day long, all through the night:** Wisława Szymborska, *Map: Collected and Last Poems*, trans. Clare Cavanagh and Stanisław Barańczak (Boston: Houghton Mifflin Harcourt, 2015).

CHAPTER 1: IDEOLOGICAL POSSESSION

11 **"political language":** George Orwell, *Why I Write*, Great Ideas 20 (London: Penguin Books, 2004).

11 **"there is a certain uniformity":** Eric Hoffer, *The True Believer: Thoughts on the Nature of Mass Movements* (New York: Harper Perennial Modern Classics, 2010), xii.

CHAPTER 2: AN EXPERIMENT

20 **my colleagues and I invited thousands of people:** For a review of the experiments and results see: Leor Zmigrod, "The Role of Cognitive Rigidity in Political Ideologies: Theory, Evidence, and Future Directions," *Current Opinion in Behavioral Sciences* 34, Political Ideologies (August 1, 2020): 34–39, https://doi.org/10.1016/j.cobeha.2019.10.016.

For the individual experiments, see: Leor Zmigrod et al., "Cognitive Flexibility and Religious Disbelief," *Psychological Research* 83, no. 8 (November 1, 2019): 1749–59, https://doi.org/10.1007/s00426-018-1034-3; Leor Zmigrod, Peter J. Rentfrow, and Trevor W. Robbins, "Cognitive Underpinnings of Nationalistic Ideology in the Context of Brexit," *Proceedings of the National Academy of Sciences* 115, no. 19 (May 8, 2018): E4532–40, https://doi.org/10.1073/pnas.1708960115; Leor Zmigrod, Peter Jason Rentfrow, and Trevor W. Robbins, "Cognitive Inflexibility Predicts Extremist Attitudes," *Frontiers in Psychology* 10 (May 7, 2019), https://doi.org/10.3389/fpsyg.2019.00989; Leor Zmigrod, Peter Jason Rentfrow, and Trevor W. Robbins, "The Partisan Mind: Is Extreme Political Partisanship Related to Cognitive Inflexibility?," *Journal of Experimental Psychology: General* 149, no. 3 (2020): 407–18, https://doi.org/10.1037/xge0000661; Leor Zmigrod et al., "The Psychological Roots of Intellectual Humility: The Role of Intelligence and Cognitive Flexibility," *Personality and Individual Differences* 141 (April 15, 2019): 200–208, https://doi.org/10.1016/j.paid.2019.01.016.

20 **Wisconsin Card Sorting Test:** The Wisconsin Card Sorting Test was invented in 1948 by researchers at the University of Wisconsin-Madison: graduate student Esta A. Berg (later Thomas) and her supervisor David Grant. It is considered a "gold standard" for measuring cognitive flexibility in humans and has often been used to assess cognitive flexibility in clinical patients and primates.

Esta A. Berg, "A Simple Objective Technique for Measuring Flexibility in Thinking," *Journal of General Psychology* 39, no. 1 (1948): 15–22, https://www.tandfonline.com/doi/abs/10.1080/00221309.1948.9918159; David A. Grant and Esta Berg, "A Behavioral Analysis of Degree of Reinforcement and Ease of Shifting to New Responses in a Weigl-Type Card-Sorting Problem," *Journal of Experimental Psychology* 38, no. 4 (1948): 404–11, https://doi.org/10.1037/h0059831.

CHAPTER 3: METAPHORS WE BELIEVE BY

23 **"We all have relatives":** Terrance Hayes, "The Art of Poetry No. 111," *Paris Review*, 2022, https://www.theparisreview.org/interviews/7930/the-art-of-poetry-no-111-terrance-hayes.

In an interview with Hilton Als, Terrance Hayes was speaking of the influence of the poet Wallace Stevens on his own writing. Hayes had written a poem, "Snow for Wallace Stevens," in which he writes: "Who is not more than his limitations, / who is not the blood in a wine barrel / and the wine as well?"

24 **"metaphors may create realities":** George Lakoff and Mark Johnson, *Metaphors We Live By* (Chicago and London: University of Chicago Press, 1980), 156.

24 **"sheer thoughtlessness":** Hannah Arendt, *Eichmann in Jerusalem: A Report on the Banality of Evil* (New York: Penguin Classics, 2006), 287.

24 **"never realized what he was doing":** Arendt, *Eichmann in Jerusalem*, 287.

24 **"total absence of thinking":** Hannah Arendt, "Thinking and Moral Considerations: A Lecture," *Social Research* 38, no. 3 (1971): 418.

24 **"extraordinary shallowness":** Arendt, "Thinking and Moral Considerations: A Lecture," 417.

24 **"inability to think":** Arendt, "Thinking and Moral Considerations: A Lecture," 417.

24 **"need no force and no persuasion":** Arendt, "Thinking and Moral Considerations: A Lecture," 436.

26 **Milgram's electrocution experiments:** Details from original study: Stanley Milgram, "Behavioral Study of Obedience," *Journal of Abnormal and Social Psychology* 67 (1963): 371–78, https://faculty.washington.edu/jdb/345/345%20Articles/Milgram.pdf.

Review of follow-up studies and interpretations: S. Alexander Haslam and Stephen D. Reicher, "50 Years of 'Obedience to Authority': From Blind Conformity to Engaged Followership," *Annual Review of Law and Social Science* 13 (October 13, 2017): 59–78, https://doi.org/10.1146/annurev-lawsocsci-110316-113710.

26 **Asch's conformity experiments:** Solomon Asch, "Effects of Group Pressure upon the Modification and Distortion of Judgments," in *Groups, Leadership and Men: Research in Human Relations*, ed. Harold Guetzkow (Pittsburgh: Carnegie Press, 1951), 177–90.

Writing about the individual differences in performance on the conformity task, Asch concludes: "One-fourth of the critical subjects was completely independent; at the other extreme, one-third of the group displaced the estimates toward the majority in one-half or more of the trials" (181–82).

26 **Zimbardo's prison experiment:** For the original experiment, see: Philip Zimbardo et al., "The Stanford Prison Experiment: A Simulation Study of the Psychology of Imprisonment" (Stanford University Press, 1971).

For one example of a critical perspective, see: Thibault Le Texier, "Debunking the Stanford Prison Experiment," *American Psychologist* 74, no. 7 (2019): 823–39, https://doi.org/10.1037/amp0000401. An exploration of the participants' various responses and rebellions to the experiment is outlined in Philip Zimbardo, *The Lucifer Effect: How Good People Turn Evil* (New York: Random House, 2011).

27 **"soul and body are two substances":** Full quote reads: "But I will say, for your benefit at least, that the whole problem contained in such questions arises simply from a supposition that is false and cannot in any way be proved, namely that, if the soul and the body are two substances whose nature is different, this prevents them from being able to act on each other." Reference for French standard edition: René Descartes, *Oeuvres de Descartes*, trans. Charles Adam and Paul Tannery, 11 vols., vol. 7 (Paris: Vrin, 1974), 213. Reference for English standard edition: René Descartes, *The Philosophical Writings of Descartes*, trans. John Cottingham, Robert Stoothoff, and Dugald Murdoch, 2 vols., vol. 2 (Cambridge University Press, 1984), 275.

28 **if the soul is intangible:** See letter from Princess Elisabeth of Bohemia to René Descartes on 6 May 1643. Princess Elisabeth of Bohemia and René Descartes, *The Correspondence Between Princess Elisabeth of Bohemia and René Descartes*, trans. Lisa Shapiro (University of Chicago Press, 2007).

28 **pineal gland:** For a discussion of Descartes's justifications for assigning the pineal gland the function of the seat of the soul, see: Lisa Shapiro, "Descartes's Pineal Gland Reconsidered," *Midwest Studies in Philosophy* 35, no. 1 (December 2011): 259–86, https://doi.org/10.1111/j.1475-4975.2011.00219.x.

CHAPTER 4: THE BIRTH OF IDEOLOGY

33 **Count Antoine Louis Claude Destutt de Tracy:** For detailed histories of Count Antoine Destutt de Tracy and his involvement in coining "ideology," please see the following resources: Terry Eagleton, *Ideology* (London: Routledge, 2014); Michael Freeden, "The 'Beginning of Ideology' Thesis," *Journal of Political Ideologies* 4, no. 1 (February 1999): 5–11, https://doi.org/10.1080/13569319908420786; Brian William Head, "Scientific Method and Ideology," in *Ideology and Social Science: Destutt de Tracy and French Liberalism* (Dordrecht: Springer Netherlands, 1985), 25–44, https://doi.org/10.1007/978-94-009-5159-4_2; Emmet Kennedy, "'Ideology' from Destutt De Tracy to Marx," *Journal of the History of Ideas* 40, no. 3 (July 1979): 353–68, https://doi.org/10.2307/2709242; George Lichtheim, "The Concept of Ideology," *History and Theory* 4, no. 2 (1965): 164–95, https://doi.org/10.2307/2504150; Manfred B. Steger, *The Rise of the Global Imaginary: Political Ideologies from the French Revolution to the Global War on Terror* (Oxford University Press, 2009), https://doi.org/10.1093/acprof:oso/9780199286942.001.0001; Bo Stråth, "Ideology and Conceptual History," in *The Oxford Handbook of Political Ideologies*, ed. Michael Freeden, Lyman Tower Sargent, and Marc Stears (Oxford: Oxford University Press, 2013), 3–19.

35 **"wipe[d] his eyes":** Anicius Manlius Severinus Boethius, chapter 2 in "The Sorrows of Boethius," book I in *The Consolation of Philosophy of Boethius*, trans. H. R. James, 5 vols., vol. 1 (London: Elliot Stock, 1897), https://www.gutenberg.org/files/14328/14328-h/14328-h.htm.

35 **"the gloom of night was scattered":** Anicius Manlius Severinus Boethius, "The Mists Dispelled," song III in "The Sorrows of Boethius," book I in *The Consolation of Philosophy of Boethius*.

35 **"Stone walls":** Richard Lovelace, "To Althea, from Prison" (1642), collected in *Lucasta* (1649), https://www.poetryfoundation.org/poems/44657/to-althea-from-prison.

36 **"The only good intellectual mechanisms":** Antoine Louis Claude, Count Destutt de Tracy, *Logique* (Logic), vol. 5, *Élémens d'idéologie* (*The Elements of Ideology*) (Paris, 1805), 52–54. Cited in Head, "Scientific Method and Ideology," 26.

36 **"leave all the other [methods]":** Tracy, *Logique* (Logic), 52–4. Cited in Head, "Scientific Method and Ideology," 26.

36 **"re-make entirely the human mind":** Cited in Head, "Scientific Method and Ideology," 26.

37 **"show the human intellect":** Tracy, *Logique* (Logic), 62. Cited in Head, "Scientific Method and Ideology," 26.

38 **pseudo-algebraic equations:** Contents of equations cited in Head, *Ideology and Social*

Science, 10. Corroborated in François Picavet's exploration of the Ideologues group: François Picavet, *Les idéologues. Essai sur l'histoire des idées et des théories scientifiques, philosophiques, religieuses, etc. en France depuis 1789*, ed. Jean-Marie Tremblay, Les classiques des sciences sociales (1891), 62, http://classiques.uqac.ca/classiques/picavet_francois/les_ideologues/ideologues.html.

In French: "Le 5 thermidor, pendant qu'on faisait l'appel des quarante-cinq condamnés qui devaient être traduits devant le tribunal révolutionnaire, il résumait la théorie à laquelle il était arrivé en formules concises: 'Le produit de la faculté de penser ou percevoir = connaissance = vérité . . . Dans un deuxième ouvrage auquel je travaille, je fais voir qu'on doit ajouter à cette équation ces trois autres membres = vertu = bonheur = sentiment d'aimer; et dans un troisième je prouverai qu'on doit ajouter ceux-ci: = liberté = égalité = philanthropie. C'est faute d'une analyze assez exacte qu'on n'est pas encore parvenu à trouver les déductions ou propositions moyennes propres à rendre palpable l'identité de ces idées.'"

40 **Pascal's famous wager:** Pascal's wager is a famous philosophical thought experiment in which Blaise Pascal considered whether one should believe in God because the costs of not doing so are substantially higher, and potentially eternal, than the costs one must incur for believing.

40 **lap he once sat in awe:** The meeting between Tracy and Voltaire took place in 1770 in Ferney, when Tracy was sixteen. This was a common rite of passage for young men of intellectual ambitions at the time. It left a strong impression on Tracy.

40 **"indecent and childish":** Voltaire, "Remarques (Premières) Sur Les Pensées de Pascal (1728)," April 18, 2012, https://web.archive.org/web/20120418162422/http://www.voltaire-integral.com/Html/22/07_Pascal.html. In French: "D'ailleurs, cet article parait un peu indécent et puéril; cette idée de jeu, de perte et de gain, ne convient point à la gravité du sujet; de plus l'intérêt que j'ai à croire une chose n'est pas une preuve de l'existence de cette chose.'"

42 **"Ideology would change the face":** Referring to the words of Maine de Biran in his letter to abbé Feletz on 30 July 1802. Cited in Emmet Kennedy, "'Ideology' from Destutt De Tracy to Marx," *Journal of the History of Ideas* 40, no. 3 (July 1979): 353–68, https://doi.org/10.2307/2709242. Original text: Maine de Biran, *Oeuvres de Maine de Biran*, ed. Pierre Tisserand, 14 vols., vol. 6 (Paris, 1922–49), 140.

In original French: "*L'idéologie*, m'ont-ils dit, doit changer la face du monde et voilà justement pourquoi ceux qui voudroient que le monde demeurât toujours bête (et pour cause) détestent l'idéologie et les idéologues." Also accessible at the following: *A.L.C. Desttut de Tracy et L'Idéologie*, Corpus Revue de Philosophie, Corpus nos. 26/27 (Paris), accessed May 17, 2024, https://revuecorpus.com/pdf/CORPUS%20N%C2%B026:27.pdf.

CHAPTER 5: THE AGE OF ILLUSIONS

44 **"Yes, they are obsessed with meddling":** Cited in Manfred B. Steger, "Ideology and Revolution: From Superscience to False Consciousness," in *The Rise of the Global Imaginary: Political Ideologies from the French Revolution to the Global War on Terror* (Oxford University Press, 2009), 19, https://doi.org/10.1093/acprof:oso/9780199286942.001.0001.

44 **"ideology" became an insult:** See Bo Stråth, "Ideology and Conceptual History," in *The Oxford Handbook of Political Ideologies*, ed. Michael Freeden, Lyman Tower Sargent, and Marc Stears (Oxford University Press, 2013), 3–19; Terry Eagleton, *Ideology* (London: Routledge, 2014), 5.

44 **"They are dreamers":** Cited in Emmet Kennedy, "'Ideology' from Destutt De Tracy to Marx," *Journal of the History of Ideas* 40, no. 3 (July 1979): 353–68, https://doi.org/10.2307/2709242.

As recalled by Prince de Talleyrand in Charles-Maurice de Talleyrand-Périgord, *Mémoires*, ed. Albert de Broglie, 5 vols., vol. 1 (Paris: Calmann Lévy, 1891), 452, http://archive.org/details/mmoiresduprince03broggoog. Full quote in French: "ce sont des rêveurs et des rêveurs dangereux; ce sont tous des matérialistes déguisés et pas trop déguisé."

44 **"It is the doctrine of the ideologues":** Cited in Eagleton, *Ideology*, 177, which refers to quote in A. Naess, *Democracy, Ideology, and Objectivity* (Oslo University Press, 1956), 151.

44 **Madame de Staël:** The French philosopher and commentator Germaine de Staël wrote about Napoleon extensively, notably in "Intoxication of Power; Reverses and Abdication of Bonaparte," chapter XIX in *Considerations on the Principal Events of the French Revolution* (1798; Indianapolis: Liberty Fund, 2008), https://oll.libertyfund.org/titles/craiutu-considerations-on-the-principal-events-of-the-french-revolution-lf-ed.

45 **"ideophobia":** Cited in Andrew Vincent, *Modern Political Ideologies* (Hoboken, NJ: John Wiley & Sons, 2009), 3.

45 **"Gentlemen, philosophers torment themselves":** Cited in Kennedy, "'Ideology' from Destutt De Tracy to Marx."

As recalled by Prince de Talleyrand in Charles Maurice de Talleyrand-Périgord, *Mémoires*, ed. Albert de Broglie, 5 vols., vol. 1 (Paris, Calmann Lévy, 1891), 452, http://archive.org/details/mmoiresduprince03broggoog. In French: "Messieurs, dit-il en élevant la voix, les philosophes se tourmentent à créer des systèmes; ils en chercheront en vain un meilleur que celui du christianisme qui, en réconciliant l'homme avec lui-même, assure en même temps l'ordre public et le repos des États. Vos idéologues détruisent toutes les illusions; et l'âge des illusions est pour les peuples comme pour les individus l'âge du bonheur."

45 **"science of lunacy," a "theory of delirium":** Written in a letter to Thomas Jefferson by John Adams—the second president of the United States—on 16 December 1816. Correspondence reproduced in Lester J. Cappon, ed., *The Adams-Jefferson Letters* (University of North Carolina Press, 1959), https://uncpress.org/book/9780807842300/the-adams-jefferson-letters/. Cited in Kennedy, "'Ideology' from Destutt De Tracy to Marx," 361.

46 **"to abolish religion as the *illusory* happiness":** Karl Marx, "A Contribution to the Critique of Hegel's Philosophy of Right: Introduction," in *Marx/Engels, Collected Works*, vol. 3 (London: Lawrence & Wishart, 1975), 175–76; more on the intellectual origins of the opium metaphor: E. O. Pedersen, "Religion Is the Opium of the People: An Investigation into the Intellectual Context of Marx's Critique of Religion," *History of Political Thought* 36, no. 2 (January 1, 2015): 354–87.

46 **"fish-blooded bourgeois doctrinaire":** Karl Marx, *Capital: A Critique of Political Economy*, trans. Ben Fowkes, 2 vols., vol. 1 (London: Penguin Books, 1976), 802.

46 **"our ideas are the necessary consequences":** Quoted in Hans Barth, *Truth and Ideology* (University of California Press, 2023), 33. Original text: Claude Adrien Helvétius, *De L'Esprit*, ed. Jean-Marie Tremblay (Paris: Durand Librairie, 1758), http://classiques.uqac.ca/classiques/helvetius_claude_adrien/de_l_esprit/de_l_esprit.html.

In French: "Il sait que nos idées sont, si je l'ose dire, des conséquences si nécessaires des sociétés où l'on vit, des lectures qu'on fait et des objets qui s'offrent à nos yeux, qu'une intelligence supérieure pourrait également, et par les objets qui se sont présentés à nous, deviner nos pensées; et, par nos pensées, deviner le nombre et l'espèce des objets que le hasard nous a offerts."

47 **"ideas of the ruling class":** Karl Marx and Friedrich Engels, "Ruling Class and Ruling Ideas," segment in "Feuerbach. Opposition of the Materialist and Idealist Outlook: Section B, The Illusion of the Epoch," in *The German Ideology*, 2 vols., vol. 1 ([1845], 1932), https://www.marxists.org/archive/marx/works/1845/german-ideology/ch01b.htm.

Full quote: "The ideas of the ruling class are in every epoch the ruling ideas, i.e. the class which is the ruling material force of society, is at the same time its ruling intellectual force."

47 **"In all ideology":** Karl Marx and Friedrich Engels, "The Essence of the Materialist Conception of History. Social Being and Social Consciousness," segment in "Feuerbach. Opposition of the Materialist and Idealist Outlook: Section A, Idealism and Materialism," in *The German Ideology*, 2 vols., vol. 1 ([1845], 1932), https://www.marxists.org/archive/marx/works/1845/german-ideology/ch01a.htm.

Full quote: "If in all ideology men and their circumstances appear upside-down as in a *camera obscura*, this phenomenon arises just as much from their historical life-process as the inversion of objects on the retina does from their physical life-process."

47 ***false consciousness:*** The clearest use of the phrase in early Marxist discourses can be found in a letter written by Friedrich Engels in 1893: Friedrich Engels to Franz Mehring, 14 July 1893, "Letters: Marx-Engels Correspondence 1893," https://www.marxists.org/archive/marx/works/1893/letters/93_07_14.htm.

For historical overviews of the term "false consciousness" in relation to ideology, see: Eagleton, *Ideology*; Ron Eyerman, "False Consciousness and Ideology in Marxist Theory," *Acta Sociologica* 24, nos. 1/2 (1981): 43–56, https://www.jstor.org/stable/4194332.

47 **"phantoms inhabiting the human brain":** Marx and Engels, "The Essence of the Materialist Conception of History. Social Being and Social Consciousness," in *The German Ideology*, https://www.marxists.org/archive/marx/works/1845/german-ideology/ch01a.htm. Sometimes translated as "phantoms formed in the human brain."

49 **"Life is not determined by consciousness":** Karl Marx wrote this in the Preface to *A Contribution to the Critique of Political Economy*, first published in 1859. Note that there are some slight differences in the phrasing according to different translators. Original German: "Es ist nicht das Bewußtsein der Menschen, das ihr Sein, sondern umgekehrt ihr gesellschaftliches Sein, das ihr Bewußtsein bestimmt."

52 **(mis)remembered as a "stream of consciousness":** In William James's 1890 *The Principles of Psychology*, he discusses the stream of thought in Chapter IX, "The Stream of Thought." He does also invoke the term "stream of consciousness," but this is less frequent than "stream of thought," and in his later work—such as his 1904 article "Does 'Consciousness' Exist?" (*Journal of Philosophy, Psychology and Scientific Methods* 1, no. 18 [September 1, 1904])—he clearly delineates "the stream of thinking (which I recognize emphatically as a phenomenon)" (491), while claiming that consciousness "is the name of a nonentity, and has no right to a place among first principles" (477).

52 **"I believe that 'consciousness'":** James, "Does 'Consciousness' Exist?," 477.

52 **"It is a peculiar sensation, this double-consciousness":** W. E. B. Du Bois, "Strivings of the Negro People," *Atlantic*, August 1897, https://www.theatlantic.com/magazine/archive/1897/08/strivings-of-the-negro-people/305446/.

CHAPTER 6: BEING A BRAIN

56 **"improve by experience":** David Hume, *An Enquiry Concerning Human Understanding* (Milton Keynes: Simon & Brown, 2011), 33.

61 **male frogs' vocalizations to attract female mates**: Bob B. M. Wong et al., "Do Temperature and Social Environment Interact to Affect Call Rate in Frogs (*Crinia signifera*)?," *Austral Ecology* 29, no. 2 (2004): 209–14, https://doi.org/10.1111/j.1442-9993.2004.01338.x; Guangzhan Fang et al., "Male Vocal Competition Is Dynamic and

Strongly Affected by Social Contexts in Music Frogs," *Animal Cognition* 17, no. 2 (March 1, 2014): 483–94, https://doi.org/10.1007/s10071-013-0680-5.

61 **"All real living is meeting":** Martin Buber, *I and Thou*, trans. Ronald Gregor Smith, Bloomsbury Revelations (London: Bloomsbury, 2013), 9.

61 **"Where there is no sharing":** Buber, *I and Thou*, 45.

62 **"collective effervescence":** Detailed in Chapter VII of Émile Durkheim, *The Elementary Forms of the Religious Life*, trans. Joseph Ward Swain (1915; Project Gutenberg, 2012), https://www.gutenberg.org/files/41360/41360-h/41360-h.htm#Page_214.

63 **the unexamined life is not worth living:** As quoted in the *Apology*, Plato's recounting of Socrates's trial. Plato, *Apology*, trans. Benjamin Jowett, https://classics.mit.edu/Plato/apology.html.

63 **"there is but one truly serious philosophical problem":** Albert Camus, *The Myth of Sisyphus*, Great Ideas 39 (London: Penguin Books, 2005), 1.

CHAPTER 7: THINKING, IDEOLOGICALLY

65 **"ideological thinking orders facts":** Hannah Arendt, *The Origins of Totalitarianism* (Milton Keynes: Benediction Classics, 2009), 471.

69 **"irresistible force of logic":** Stalin's speech of 28 January 1924, in which he reflected on Vladimir Lenin's power as an orator and leader. Quoted from Lenin, *Selected Works*, vol. 1 (Moscow: Progress Publishers, 1947), 33. Cited in Arendt, *The Origins of Totalitarianism.*

69 **"ideologies are sealed universes":** Terry Eagleton, *Ideology* (London: Routledge, 2014), 10.

71 **"genealogically lucky":** Amia Srinivasan, "VII—Genealogy, Epistemology and Worldmaking," *Proceedings of the Aristotelian Society* 119, no. 2 (July 1, 2019): 127–56, https://doi.org/10.1093/arisoc/aoz009.

72 **Both the doctrinal and relational dimensions are necessary:** For a more academic discussion of this, see: Leor Zmigrod, "A Psychology of Ideology: Unpacking the Psychological Structure of Ideological Thinking," *Perspectives on Psychological Science* 17, no. 4 (July 1, 2022): 1072–92, https://doi.org/10.1177/17456916211044140.

76 **"we might as well give up":** Steven Pinker, *Enlightenment Now* (London: Allen Lane, 2018), 353.

77 **"prejudice is unlikely to be merely a specific attitude":** Gordon W. Allport, *The Nature of Prejudice* (Cambridge, MA: Addison-Wesley, 1954), 175, http://archive.org/details/in.ernet.dli.2015.188638.

CHAPTER 8: A CHICKEN-AND-EGG PROBLEM

83 **dramatized sketches of real interviews:** In every vignette of interviewed children speaking throughout the book, the speech is directly quoted from the interview materials and results reported by Else Frenkel-Brunswik in academic papers and archival material. Sometimes a vignette may follow the reasoning of a single participant, and sometimes a vignette may incorporate the reports of multiple participants who were asked the same questions. This is done in order to give the reader a coherent sense of the interviews and responses and to distill at times fragmentary results reported in the original papers. The character of the interviewer in these vignettes is a creative interpretation that seeks to capture Frenkel-Brunswik's broader writings and reflections, as well as the methodology. In the original interviews, there were several interviewers at work with different participants—these included Frenkel-Brunswik's colleagues, students, and assistants. Joan Havel (Grant), a graduate student at Berkeley, was heavily involved in the interview methodology and was later named as one of the executors of Frenkel-Brunswik's estate and archival materials. Frenkel-Brunswik also named Murray Jarvik and Milton Rokeach as notable collaborators in designing tests of prejudice and cognition.

83 **"ethnocentric child becomes a potential fascist":** Else Frenkel-Brunswik, "A Study of Prejudice in Children," *Human Relations* 1 (1948): 295–306, https//doi.org/10.1177/001872674800100301.

83 **California children into experimental participants:** The studies were conducted at the Institute of Child Welfare of the University of California. For more details, see Frenkel-Brunswik, "A Study of Prejudice in Children"; Else Frenkel-Brunswik and Joan Havel, "Prejudice in the Interviews of Children. I. Attitudes toward Minority Groups," *Journal of Genetic Psychology* 82, no. 1 (March 1953): 91–136.

84 **"If a potentially fascistic individual exists":** T. W. Adorno et al., *The Authoritarian Personality*, Studies in Prejudice (New York: Harper, 1950), 2.

85 **"treat the brain like a muscle":** "Treating the Brain Like a Muscle, Not a Sponge," St. Mary's School, Cambridge, https://www.stmaryscambridge.co.uk/news-and-blog/blog/view~treating-the-brain-like-a-muscle-not-a-sponge_8994.htm.

87 **chicken-and-egg problem of political neuroscience:** Conceptual and methodological discussions of the chicken-and-egg problem of political neuroscience can be found at the following: Leor Zmigrod, "A Neurocognitive Model of Ideological Thinking," *Politics and the Life Sciences* 40, no. 2 (October 2021): 224–38, https://doi.org/10.1017/pls.2021.10; Ingrid J. Haas, Clarisse Warren, and Samantha J. Lauf, "Political Neuroscience: Understanding How the Brain Makes Political Decisions," in *Oxford Research Encyclopedia of Politics*, 2020, https://doi.org/10.1093/acrefore/9780190228637.013.948; John T. Jost et al., "Political Neuroscience: The Beginning of a Beautiful Friendship," *Political Psychology* 35, no. S1 (2014): 3–42, https://doi.org/10.1111/pops.12162; John T. Jost, *Left and Right: The Psychological Significance of a Political Distinction* (Oxford University

Press, 2021); Hyun Hannah Nam, "Neuroscientific Approaches to the Study of System Justification," in "Political Ideologies," ed. John Jost, Eran Halperin, and Kristin Laurin, special issue, *Current Opinion in Behavioral Sciences* 34 (August 1, 2020): 205–10, https://doi.org/10.1016/j.cobeha.2020.04.003; Peter Beattie, "The 'Chicken-and-Egg' Development of Political Opinions: The Roles of Genes, Social Status, Ideology, and Information," *Politics and the Life Sciences* 36, no. 1 (April 2017): 1–13, https://doi.org/10.1017/pls.2017.1.

88 **"what totalitarian ideologies aim at":** Hannah Arendt, *The Origins of Totalitarianism* (Milton Keynes: Benediction Classics, 2009), 458.

CHAPTER 9: YOUNG AUTHORITARIANS

89 **As a Jewish child in Vienna:** There are several historical resources on Else Frenkel-Brunswik's life. A few useful summaries include:

The special issue outlined in: Andreas Kranebitter and Christoph Reinprecht, "Authoritarianism, Ambivalence, Ambiguity: The Life and Work of Else Frenkel-Brunswik. Introduction to the Special Issue," *Serendipities. Journal for the Sociology and History of the Social Sciences* 7, nos. 1–2 (January 10, 2023): 1–12, https://doi.org/10.7146/serendipities.v7i1-2.135380.

Jaan Valsiner and Emily Abbey, "Ambivalence in Focus: Remembering the Life and Work of Else Frenkel-Brunswik," *Studies in Psychology* 27, no. 1 (January 1, 2006): 9–17, https://doi.org/10.1174/021093906776173126.

90 **"among the most powerless":** Merve Emre, review of *Making It Big*, by Matthew Dennison, *New York Review of Books*, December 22, 2022, https://www.nybooks.com/articles/2022/12/22/making-it-big-roald-dahl-teller-of-the-unexpected/.

91 **"more direct and uninhibited":** Else Frenkel-Brunswik and Joan Havel, "Prejudice in the Interviews of Children. I. Attitudes toward Minority Groups," *Journal of Genetic Psychology* 82, no. 1 (March 1953): 92, https://doi.org/10.1080/08856559.1953.10533657.

94 **"The absoluteness of each of these differences":** Else Frenkel-Brunswik, "Intolerance of Ambiguity as an Emotional and Perceptual Personality Variable," *Journal of Personality* 18, no. 1 (1949): 117, https://doi.org/10.1111/j.1467-6494.1949.tb01236.x.

CHAPTER 10: BRAINWASHING A BABY

97 **"experiments of living":** John Stuart Mill, *On Liberty*, Great Ideas 86 (London: Penguin Books, 2010), 82.

101 **"Attention consists of suspending our thought":** Simone Weil, "Reflections on the Right Use of School Studies with a View to the Love of God," in *Waiting for God*, trans. Emma Craufurd (New York: Harper Perennial, 1973), 111, https://antilogicalism.com/wp-content/uploads/2019/04/waiting-god.pdf.

104 **"We may conceive of the story":** Else Frenkel-Brunswik, "Intolerance of Ambiguity as an Emotional and Perceptual Personality Variable," *Journal of Personality* 18, no. 1 (1949): 124, https://doi.org/10.1111/j.1467-6494.1949.tb01236.x.

104 **"negativistic tendencies in the distortion":** Frenkel-Brunswik, "Intolerance of Ambiguity as an Emotional and Perceptual Personality Variable," 124.

104 **"In the recall of the prejudiced children":** Frenkel-Brunswik, "Intolerance of Ambiguity as an Emotional and Perceptual Personality Variable," 124.

104 **"stray from the content of the story":** Frenkel-Brunswik, "Intolerance of Ambiguity as an Emotional and Perceptual Personality Variable," 126.

104 **"both these patterns help avoidance of uncertainty":** Frenkel-Brunswik, "Intolerance of Ambiguity as an Emotional and Perceptual Personality Variable," 126.

106 **"It is as if any stimulus":** Frenkel-Brunswik, "Intolerance of Ambiguity as an Emotional and Perceptual Personality Variable," 128–29.

CHAPTER 11: THE RIGID MIND

112 **After inviting thousands of participants:** Leor Zmigrod et al., "The Psychological Roots of Intellectual Humility: The Role of Intelligence and Cognitive Flexibility," *Personality and Individual Differences* 141 (April 15, 2019): 200–208, https://doi.org/10.1016/j.paid.2019.01.016.

113 **"What art and morality have in common":** Jean-Paul Sartre, *Existentialism Is a Humanism* (New Haven, CT: Yale University Press, 2007), 46.

113 **"Has anyone ever blamed an artist":** Sartre, *Existentialism Is a Humanism*, 45.

114 **Jaensch created a personality typology:** Outlined in E. R. Jaensch's 1938 book *Der Gegentypus*, which may be translated as the "Anti-Type" or the "Counter-Type."

114 **Jaensch claimed S-types were probably communist:** More details on Jaensch's framing can be found in: Roger Brown, "The Authoritarian Personality and the Organization of Attitudes," in *Political Psychology: Key Readings*, ed. John T. Jost and Jim Sidanius (New York: Psychology Press, 2004).

114 **Frenkel-Brunswik observed that "the sampling technique":** Else Frenkel-Brunswik, "Intolerance of Ambiguity as an Emotional and Perceptual Personality Variable," *Journal of Personality* 18, no. 1 (1949): 111–12, https://doi.org/10.1111/j.1467-6494.1949.tb01236.x.

For more on the intersection between Frenkel-Brunswik and Jaensch's work, see: Andreas Kranebitter and Fabian Gruber, "Allowing for Ambiguity in the Social Sciences: Else Frenkel-Brunswik's Methodological Practice in *The Authoritarian Personality*,"

Serendipities. Journal for the Sociology and History of the Social Sciences 7, nos. 1–2 (January 10, 2023): 30–59, https://doi.org/10.7146/serendipities.v7i1-2.132541.

115 **"citizen of nowhere":** As Theresa May claimed at the Conservative Party Conference in 2016. See "Full Text: Theresa May's Conference Speech," *Spectator*, October 5, 2016, https://www.spectator.co.uk/article/full-text-theresa-may-s-conference-speech/.

117 **"What is love of one's country":** Ursula K. Le Guin, "To the Ice," chapter 15 in *The Left Hand of Darkness* (New York: Ace Books, 1969).

117 **hypothesized that nationalistic thinking:** Leor Zmigrod, Peter J. Rentfrow, and Trevor W. Robbins, "Cognitive Underpinnings of Nationalistic Ideology in the Context of Brexit," *Proceedings of the National Academy of Sciences* 115, no. 19 (May 8, 2018): E4532–40, https://doi.org/10.1073/pnas.1708960115.

119 **assumption that the political right is naturally rigid:** One of the most notable and expanded versions of this hypothesis is encapsulated in: John T. Jost et al., "Political Conservatism as Motivated Social Cognition," *Psychological Bulletin* 129, no. 3 (2003): 339–75, https://doi.org/10.1037/0033-2909.129.3.339.

119 **left-wing believer may not like to see themselves:** For the role of self-perceptions, see: Bethany Lassetter and Rebecca Neel, "Malleable Liberals and Fixed Conservatives? Political Orientation Shapes Perceived Ability to Change," *Journal of Experimental Social Psychology* 82 (May 1, 2019): 141–51, https://doi.org/10.1016/j.jesp.2019.01.002.

120 **cognitive and behavioral approaches:** For a comparison of the evidence for the rigidity-of-the-right hypothesis with self-report measures of flexibility versus behavioral measures, see: Alain Van Hiel et al., "The Relationship Between Right-Wing Attitudes and Cognitive Style: A Comparison of Self-Report and Behavioural Measures of Rigidity and Intolerance of Ambiguity," *European Journal of Personality* 30, no. 6 (November 1, 2016): 523–31, https://doi.org/10.1002/per.2082.

120 **Antonio Gramsci ridiculed the idea:** Gramsci wrote of this in his *Prison Notebooks*, in particular on p. 116 of the first volume (out of six), as cited in: Michele Filippini, "Ideology," in *Using Gramsci: A New Approach*, trans. Patrick J. Barr (London: Pluto Press, 2017), 10, https://www.jstor.org/stable/j.ctt1h64kxd.7.

120 **Philip Converse conveyed:** Philip E. Converse, "The Nature of Belief Systems in Mass Publics (1964)," *Critical Review* 18, nos. 1–3 (January 2006): 1, https://doi.org/10.1080/08913810608443650.

121 **adopted a strikingly different policy position:** Gustavo A. Flores-Macías, "Statist vs. Pro-Market: Explaining Leftist Governments' Economic Policies in Latin America," *Comparative Politics* 42, no. 4 (July 1, 2010): 413–33, https://doi.org/10.5129/001041510X12911363510033.

121 **pattern was reversed:** Margit Tavits and Natalia Letki, "When Left Is Right: Party Ideology and Policy in Post-Communist Europe," *American Political Science Review* 103, no. 4 (November 2009): 555–69, https://doi.org/10.1017/S0003055409990220.

121 **traits that prompt:** Christopher M. Federico and Ariel Malka, "The Contingent, Contextual Nature of the Relationship Between Needs for Security and Certainty and Political Preferences: Evidence and Implications," *Political Psychology* 39, no. S1 (2018): 3–48, https://doi.org/10.1111/pops.12477; Ariel Malka et al., "Do Needs for Security and Certainty Predict Cultural and Economic Conservatism? A Cross-National Analysis.," *Journal of Personality and Social Psychology* 106, no. 6 (2014): 1031–51, https://doi.org/10.1037/a0036170.

123 **identity fusion as a construct and theory:** For reviews on identity fusion as a construct and theory, see: Ángel Gómez et al., "Recent Advances, Misconceptions, Untested Assumptions, and Future Research Agenda for Identity Fusion Theory," *Social and Personality Psychology Compass* 14, no. 6 (2020): e12531, https://doi.org/10.1111/spc3.12531; William B. Swann and Michael D. Buhrmester, "Identity Fusion," *Current Directions in Psychological Science* 24, no. 1 (February 1, 2015): 52–57, https://doi.org/10.1177/0963721414551363.

123 **visual measure of identity fusion:** Juan Jiménez et al., "The Dynamic Identity Fusion Index: A New Continuous Measure of Identity Fusion for Web-Based Questionnaires," *Social Science Computer Review* 34, no. 2 (April 1, 2016): 215–28, https://doi.org/10.1177/0894439314566178.

123 **In a study with over 700 Americans:** Leor Zmigrod, Peter Jason Rentfrow, and Trevor W. Robbins, "The Partisan Mind: Is Extreme Political Partisanship Related to Cognitive Inflexibility?," *Journal of Experimental Psychology: General* 149, no. 3 (2020): 407–18, https://doi.org/10.1037/xge0000661.

125 **famous philosophical conundrum known as the trolley problem:** For more examples of this intergroup trolley dilemma in the context of identity fusion, see: William B. Swann Jr. et al., "Contemplating the Ultimate Sacrifice: Identity Fusion Channels Pro-Group Affect, Cognition, and Moral Decision Making," *Journal of Personality and Social Psychology* 106, no. 5 (2014): 713–27, https://doi.org/10.1037/a0035809; William B. Swann et al., "Dying and Killing for One's Group: Identity Fusion Moderates Responses to Intergroup Versions of the Trolley Problem," *Psychological Science* 21, no. 8 (August 1, 2010): 1176–83, https://doi.org/10.1177/0956797610376656.

126 **A thousand people answered this moral dilemma:** Leor Zmigrod, Peter Jason Rentfrow, and Trevor W. Robbins, "Cognitive Inflexibility Predicts Extremist Attitudes," *Frontiers in Psychology* 10 (May 7, 2019), https://doi.org/10.3389/fpsyg.2019.00989.

CHAPTER 12: THE DOGMATIC GENE

129 **discovered that the most rigid individuals possess specific genes:** Leor Zmigrod and Trevor W. Robbins, "Dopamine, Cognitive Flexibility, and IQ: Epistatic Catechol-O-MethylTransferase:DRD2 Gene–Gene Interactions Modulate Mental Rigidity," *Journal of Cognitive Neuroscience* 34, no. 1 (December 1, 2021): 153–79, https://doi.org/10.1162/jocn_a_01784.

130 **dopamine receptors can instigate:** Cristina Missale et al., "Dopamine Receptors: From Structure to Function," *Physiological Reviews* 78, no. 1 (January 1, 1998): 189–226; Jean E. Lachowicz and David R. Sibley, "Molecular Characteristics of Mammalian Dopamine Receptors," *Pharmacology & Toxicology* 81, no. 3 (1997): 105–13, https://doi.org/10.1111/j.1600-0773.1997.tb00039.x; C. R. Yang and J. K. Seamans, "Dopamine D1 Receptor Actions in Layers V–VI Rat Prefrontal Cortex Neurons in Vitro: Modulation of Dendritic-Somatic Signal Integration," *Journal of Neuroscience* 16, no. 5 (March 1, 1996): 1922–35, https://doi.org/10.1523/JNEUROSCI.16-05-01922.1996.

131 **D_2 receptors are concentrated:** Louis-Eric Trudeau et al., "Chapter 6—The Multilingual Nature of Dopamine Neurons," in *Progress in Brain Research*, ed. Marco Diana, Gaetano Di Chiara, and Pierfranco Spano, vol. 211, Dopamine (Amsterdam: Elsevier, 2014), 141–64, https://doi.org/10.1016/B978-0-444-63425-2.00006-4.

131 **230,000 and 430,000 dopamine neurons:** Yaping Chu et al., "Age-Related Decreases in Nurr1 Immunoreactivity in the Human Substantia Nigra," *Journal of Comparative Neurology* 450, no. 3 (2002): 203–14, https://doi.org/10.1002/cne.10261.

131 **up to four meters:** J. Paul Bolam and Eleftheria K. Pissadaki, "Living on the Edge with Too Many Mouths to Feed: Why Dopamine Neurons Die," *Movement Disorders* 27, no. 12 (2012): 1478–83, https://doi.org/10.1002/mds.25135.

131 **can affect tens of thousands:** Nicolas X. Tritsch and Bernardo L. Sabatini, "Dopaminergic Modulation of Synaptic Transmission in Cortex and Striatum," *Neuron* 76, no. 1 (October 2012): 33–50, https://doi.org/10.1016/j.neuron.2012.09.023; Wakoto Matsuda et al., "Single Nigrostriatal Dopaminergic Neurons Form Widely Spread and Highly Dense Axonal Arborizations in the Neostriatum," *Journal of Neuroscience* 29, no. 2 (January 14, 2009): 444–53, https://doi.org/10.1523/JNEUROSCI.4029-08.2009.

132 **hypothesized that individual differences:** Roshan Cools, "Chemistry of the Adaptive Mind: Lessons from Dopamine," *Neuron* 104, no. 1 (October 2019): 113–31, https://doi.org/10.1016/j.neuron.2019.09.035; Roshan Cools and Mark D'Esposito, "Inverted-U–Shaped Dopamine Actions on Human Working Memory and Cognitive Control," in "Prefrontal Cortical Circuits Regulating Attention, Behavior and Emotion," special issue, *Biological Psychiatry* 69, no. 12 (June 15, 2011): e113–25, https://doi.org/10.1016/j.biopsych.2011.03.028.

132 **Experiments with "knockout" mice:** For a review see: Marianne Klanker, Matthijs Feenstra, and Damiaan Denys, "Dopaminergic Control of Cognitive Flexibility in Humans and Animals," *Frontiers in Neuroscience* 7 (November 5, 2013), https://doi.org/10.3389/fnins.2013.00201.

132 **Such knockout mice:** J. W. Smith et al., "Dopamine D2L Receptor Knockout Mice Display Deficits in Positive and Negative Reinforcing Properties of Morphine and in Avoidance Learning," *Neuroscience* 113, no. 4 (September 10, 2002): 755–65, https://doi.org/10.1016/S0306-4522(02)00257-9.

133 **differences in our taste perception:** Elie Chamoun et al., "A Review of the Associations Between Single Nucleotide Polymorphisms in Taste Receptors, Eating Behaviors, and Health," *Critical Reviews in Food Science and Nutrition* 58, no. 2 (January 22, 2018): 194–207, https://www.tandfonline.com/doi/abs/10.1080/10408398.2016.1152229; Alexey A. Fushan et al., "Allelic Polymorphism within the TAS1R3 Promoter Is Associated with Human Taste Sensitivity to Sucrose," *Current Biology* 19, no. 15 (August 2009): 1288–93, https://doi.org/10.1016/j.cub.2009.06.015.

133 **less-sensitive discriminators, or even *nontasters*:** Gretchen L. Goldstein, Henryk Daun, and Beverly J. Tepper, "Adiposity in Middle-Aged Women Is Associated with Genetic Taste Blindness to 6-n-Propylthiouracil," *Obesity Research* 13, no. 6 (2005): 1017–23, https://doi.org/10.1038/oby.2005.119.

133 **degree to which we enjoy:** Adam Drewnowski et al., "Taste and Food Preferences as Predictors of Dietary Practices in Young Women," *Public Health Nutrition* 2, no. 4 (April 1999): 513–9, https://doi.org/10.1017/S1368980099000695.

133 **the COMT gene helps to regulate levels of dopamine:** Leonid Yavich et al., "Site-Specific Role of Catechol-O-Methyltransferase in Dopamine Overflow within Prefrontal Cortex and Dorsal Striatum," *Journal of Neuroscience* 27, no. 38 (September 19, 2007): 10196–209, https://doi.org/10.1523/JNEUROSCI.0665-07.2007.

134 ***Val* allele has four times:** Herbert M. Lachman et al., "Human Catechol-O-Methyltransferase Pharmacogenetics: Description of a Functional Polymorphism and Its Potential Application to Neuropsychiatric Disorders," *Pharmacogenetics and Genomics* 6, no. 3 (June 1996): 243–50, https://journals.lww.com/jpharmacogenetics/abstract/1996/06000/human_catechol_o_methyltransferase.7.aspx; Jingshan Chen et al., "Functional Analysis of Genetic Variation in Catechol-O-Methyltransferase (COMT): Effects on mRNA, Protein, and Enzyme Activity in Postmortem Human Brain," *American Journal of Human Genetics* 75, no. 5 (November 2004): 807–21, https://doi.org/10.1086/425589.

134 ***Met*-carriers tend to perform better:** Daniela Mier, Peter Kirsch, and Andreas Meyer-Lindenberg, "Neural Substrates of Pleiotropic Action of Genetic Variation in

COMT: A Meta-Analysis," *Molecular Psychiatry* 15 (June 1, 2009): 918–27, https://doi.org/10.1038/mp.2009.36; Jonathan Flint and Marcus R. Munafò, "The Endophenotype Concept in Psychiatric Genetics," *Psychological Medicine* 37, no. 2 (February 2007): 163–80, https://doi.org/10.1017/S0033291706008750; J. H. Barnett et al., "Effects of the Catechol-O-Methyltransferase ValMet Polymorphism on Executive Function: A Meta-Analysis of the Wisconsin Card Sort Test in Schizophrenia and Healthy Controls," *Molecular Psychiatry* 12, no. 5 (May 2007): 502–9, https://doi.org/10.1038/sj.mp.4001973.

135 **in the striatum:** Or its homologue in monkeys and rodents.

136 **Finnish research group:** M. Hirvonen et al., "C957T Polymorphism of the Dopamine D2 Receptor (DRD2) Gene Affects Striatal DRD2 Availability in Vivo," *Molecular Psychiatry* 9, no. 12 (December 2004): 1060–61, https://doi.org/10.1038/sj.mp.4001561; Mika M. Hirvonen et al., "C957T Polymorphism of the Human Dopamine D2 Receptor Gene Predicts Extrastriatal Dopamine Receptor Availability *in Vivo*," *Progress in Neuro-Psychopharmacology and Biological Psychiatry* 33, no. 4 (June 15, 2009): 630–36, https://doi.org/10.1016/j.pnpbp.2009.02.021.

136 **genetic profile puts people at risk:** A review on the genetics of obsessive-compulsive disorder recently pointed to the COMT and DRD2 genotypes as risk factors for obsessive-compulsive symptoms. Future research will need to replicate these patterns in additional and larger samples and consider the involvement of other genes, such as glutamate transporter genes.

For research on the link between rigidity in obsessive-compulsive symptoms and the COMT gene and DRD2 gene, see: Kim M. Schindler et al., "Association between Homozygosity at the COMT Gene Locus and Obsessive Compulsive Disorder," *American Journal of Medical Genetics* 96, no. 6 (2000): 721–24, https://doi.org/10.1002/1096-8628(20001204)96:6<721::AID-AJMG4>3.0.CO;2-M; David L. Pauls, "The Genetics of Obsessive-Compulsive Disorder: A Review," *Dialogues in Clinical Neuroscience* 12, no. 2 (June 30, 2010): 149–63, https://doi.org/10.31887/DCNS.2010.12.2/dpauls; Humberto Nicolini et al., "DRD2, DRD3 and 5HT2A Receptor Genes Polymorphisms in Obsessive-Compulsive Disorder," *Molecular Psychiatry* 1 (January 1, 1997): 461–65; Maria Karayiorgou et al., "Genotype Determining Low Catechol-O-Methyltransferase Activity as a Risk Factor for Obsessive-Compulsive Disorder," *Proceedings of the National Academy of Sciences* 94, no. 9 (April 29, 1997): 4572–75, https://doi.org/10.1073/pnas.94.9.4572; Maria Karayiorgou et al., "Family-Based Association Studies Support a Sexually Dimorphic Effect of *COMT* and *MAOA* on Genetic Susceptibility to Obsessive-Compulsive Disorder," *Biological Psychiatry* 45, no. 9 (May 1, 1999): 1178–89, https://doi.org/10.1016/S0006-3223(98)00319-9; E. A. Billett et al., "Investigation of Dopamine System Genes in Obsessive-Compulsive Disorder," *Psychiatric Genetics* 8, no. 3 (Autumn 1998): 163–70, https://journals

.lww.com/psychgenetics/abstract/1998/00830/Investigation_of_dopamine_system_genes_in.5.aspx.

139 **traits that are not visible in parents:** Michael D. Hunter, Kevin L. McKee, and Eric Turkheimer, "Simulated Nonlinear Genetic and Environmental Dynamics of Complex Traits," *Development and Psychopathology* 35, no. 2 (May 2023): 662–77, https://doi.org/10.1017/S0954579421001796; David T. Lykken et al., "Emergenesis: Genetic Traits that May Not Run in Families," *American Psychologist* 47, no. 12 (1992): 1565–77, https://doi.org/10.1037/0003-066X.47.12.1565.

CHAPTER 13: DARWIN'S SECRET

145 **section she had described to Charles:** Emma Wedgwood to Charles Darwin, November 2, 1838 (letter no. 441), Darwin Correspondence Project, https://www.darwinproject.ac.uk/letter/?docId=letters/DCP-LETT-441.xml.

146 **"a painful void between us":** Emma Wedgwood to Charles Darwin, November 2, 1838.

146 **"May not the habit in scientific pursuits":** Emma Darwin to Charles Darwin, c. February 1839 (letter no. 471), Darwin Correspondence Project, https://www.darwinproject.ac.uk/letter/?docId=letters/DCP-LETT-471.xml.

147 **lying awake at the cusp of a yellow dawn:** In letters and recollections, including Francis Darwin's "Reminiscences of My Father's Everyday Life," Darwin's children recalled him being a terrible, troubled sleeper and that this was his typical sleeping position.

147 **"overlook the probability of the constant inculcation":** *The Autobiography of Charles Darwin*, ed. Nora Barlow (1958), 93, https://darwin-online.org.uk/content/frameset?pageseq=1&itemID=F1497&viewtype=side. Original handwritten notes available on page 73 of C. R. Darwin, "Recollections of the Development of My Mind and Character," CUL-DAR26.1-121, John van Wyhe, ed., Darwin Online, http://darwin-online.org.uk/.

147 **letter from Emma:** *Autobiography of Charles Darwin*, ed. Barlow, 93–94, note.

149 **letters to his son George:** Charles Darwin to G. H. Darwin, 21 October [1873] (letter no. 9105), Darwin Correspondence Project, https://www.darwinproject.ac.uk/letter/?docId=letters/DCP-LETT-9105.xml.

152 **cognitive flexibility was linked to religious disbelief:** Leor Zmigrod et al., "Cognitive Flexibility and Religious Disbelief," *Psychological Research* 83, no. 8 (November 1, 2019): 1749–59, https://doi.org/10.1007/s00426-018-1034-3.

154 **"Conversion tends to be a cure":** Adam Phillips, *On Wanting to Change* (London: Penguin Books, 2021), 9.

155 **"however strong men's propensity to believe invisible":** David Hume, "Various

Forms of Polytheism," section 5 in *The Natural History of Religion* (London: A. and H. Bradlaugh Bonner, 1889).

155 **"In my Journal I wrote":** *Autobiography of Charles Darwin*, ed. Barlow, 91–92.

CHAPTER 14: POLIPTICAL ILLUSIONS

157 **The German cartoonist:** For a detailed history of the duck-rabbit figure, from its initial publication to its uptake by Joseph Jastrow—one of the first American psychologists—see: John F. Kihlstrom, "Joseph Jastrow and His Duck—or Is It a Rabbit?," 2004, https://www.ocf.berkeley.edu/~jfkihlstrom/JastrowDuck.htm.

158 **switch between the faces of the duck-rabbit illusion:** Richard Wiseman et al., "Creativity and Ease of Ambiguous Figural Reversal," *British Journal of Psychology* 102, no. 3 (2011): 615–22, https://doi.org/10.1111/j.2044-8295.2011.02031.x.

The pattern was replicated in relation to AUT and the Necker cube illusion by: Annabel Blake and Stephen Palmisano, "Divergent Thinking Influences the Perception of Ambiguous Visual Illusions," *Perception* 50, no. 5 (May 1, 2021): 418–37, https://doi.org/10.1177/03010066211000192.

158 **study of over 500 Swiss participants:** Peter Brugger and Susanne Brugger, "The Easter Bunny in October: Is It Disguised as a Duck?," *Perceptual and Motor Skills* 76, no. 2 (April 1, 1993): 577–78, https://doi.org/10.2466/pms.1993.76.2.577.

159 **"a thing is funny," George Orwell observed**: George Orwell, "Funny, but Not Vulgar," *Leader*, July 28, 1945.

160 **"what is different: my impression?":** Ludwig Wittgenstein, *Philosophical Investigations*, trans. G. E. M. Anscombe (Oxford: Basil Blackwell, 1968), Part II, xi, 195.

160 **"I contemplate a face, and then suddenly notice":** Wittgenstein, *Philosophical Investigations*, Part II, xi, 193.

163 **dataset of over 300 American participants:** Leor Zmigrod et al., "The Cognitive and Perceptual Correlates of Ideological Attitudes: A Data-Driven Approach," *Philosophical Transactions of the Royal Society B: Biological Sciences* 376, no. 1822 (February 22, 2021): 20200424, https://doi.org/10.1098/rstb.2020.0424.

164 **J. Richard Simon discovered in the 1960s:** J. Richard Simon and Alan P. Rudell, "Auditory S-R Compatibility: The Effect of an Irrelevant Cue on Information Processing," *Journal of Applied Psychology* 51, no. 3 (1967): 300–304, https://doi.org/10.1037/h0020586.

164 **being distracted by irrelevant features can inspire:** Sharon Zmigrod, Leor Zmigrod, and Bernhard Hommel, "Zooming into Creativity: Individual Differences in Attentional

Global-Local Biases Are Linked to Creative Thinking," *Frontiers in Psychology* 6 (October 30, 2015), https://doi.org/10.3389/fpsyg.2015.01647.

167 **other research showing that people with right-wing views:** For example: Benjamin C. Ruisch, Natalie J. Shook, and Russell H. Fazio, "Of Unbiased Beans and Slanted Stocks: Neutral Stimuli Reveal the Fundamental Relation Between Political Ideology and Exploratory Behaviour," *British Journal of Psychology* 112, no. 1 (2021): 358–61, https://doi.org/10.1111/bjop.12455; Natalie J. Shook and Russell H. Fazio, "Political Ideology, Exploration of Novel Stimuli, and Attitude Formation," *Journal of Experimental Social Psychology* 45, no. 4 (July 2009): 995–98, https://doi.org/10.1016/j.jesp.2009.04.003.

169 **radical thinkers struggle to judge their own mental processes:** Max Rollwage, Raymond J. Dolan, and Stephen M. Fleming, "Metacognitive Failure as a Feature of Those Holding Radical Beliefs," *Current Biology* 28, no. 24 (December 2018): 4014–21.e8, https://doi.org/10.1016/j.cub.2018.10.053.

169 **"Ambiguity—rabbit or duck?":** Ernst Hans Gombrich, *Art and Illusion: A Study in the Psychology of Pictorial Representation* (Princeton University Press, 1972), http://archive.org/details/artillusionstud00gomb.

170 **"The modern style of interpretation":** Susan Sontag, *Against Interpretation and Other Essays* (London: Penguin Modern Classics, 2009), 6–7.

171 **"To interpret is to impoverish":** Sontag, *Against Interpretation and Other Essays*, 7.

171 **"the luminousness of the thing":** Sontag, *Against Interpretation and Other Essays*, 13.

171 **"What is important now":** Sontag, *Against Interpretation and Other Essays*, 14.

171 **"the revenge of the intellect":** Sontag, *Against Interpretation and Other Essays*, 7.

CHAPTER 15: YOUR EMOTIONAL FINGERTIPS

173 **"Feelings let us *mind the body*":** Antonio Damasio, *Descartes' Error: Emotion, Reason and the Human Brain* (Rochester: Vintage Digital, 2008), 123.

173 **Nebraskan researchers exposed participants to abrupt noises:** Douglas R. Oxley et al., "Political Attitudes Vary with Physiological Traits," *Science* 321, no. 5896 (September 19, 2008): 1667–70, https://doi.org/10.1126/science.1157627.

175 **more authors named John:** Issues of gender representation are at times acute in political neuroscience. One of the first political neuroscientific investigations was an all-male research team who decided to study only male participants because of supposed general differences in how men and women process emotions. Increasing the

number of women in science is critical if such prejudices and blind spots are to be avoided.

Drew Westen et al., "Neural Bases of Motivated Reasoning: An fMRI Study of Emotional Constraints on Partisan Political Judgment in the 2004 U.S. Presidential Election," *Journal of Cognitive Neuroscience* 18, no. 11 (November 1, 2006): 1947–58, https://doi.org/10.1162/jocn.2006.18.11.1947.

175 **Multiple international research teams:** As a general note, for experimental studies with more than one or two authors, the details of the researchers conducting the studies can be found in the notes and are otherwise referred to in relation to the city of the university in which the experiment was completed. This approach is adopted for concision and because acknowledging large research teams in science can be tricky: the relative contributions of different team members are not always clearly laid out. Moreover, by using the city as a proxy, this gives the reader a sense of the geographical location or diversity of the participants and the political environment in which they are based. This helps to contextualize findings, for example, that compare Dutch participants in a political system with over ten parties and American participants who are typically surveyed about the two dominant political parties.

175 **entire journal issue was dedicated:** See target article and associated commentaries for John R. Hibbing, Kevin B. Smith, and John R. Alford, "Differences in Negativity Bias Underlie Variations in Political Ideology," *Behavioral and Brain Sciences* 37, no. 3 (June 2014): 297–350, https://doi.org/10.1017/S0140525X13001192.

176 **collection of experiments supports the hypothesis:** For example: Michael D. Dodd et al., "The Political Left Rolls with the Good and the Political Right Confronts the Bad: Connecting Physiology and Cognition to Preferences," *Philosophical Transactions of the Royal Society B: Biological Sciences* 367, no. 1589 (March 5, 2012): 640–49, https://doi.org/10.1098/rstb.2011.0268. For a review, see: Kevin B. Smith and Clarisse Warren, "Physiology Predicts Ideology. Or Does It? The Current State of Political Psychophysiology Research," in "Political Ideologies," special issue, *Current Opinion in Behavioral Sciences* 34 (August 1, 2020): 88–93, https://doi.org/10.1016/j.cobeha.2020.01.001.

176 **interpret them as threatening:** Jacob M. Vigil, "Political Leanings Vary with Facial Expression Processing and Psychosocial Functioning," *Group Processes & Intergroup Relations* 13, no. 5 (September 1, 2010): 547–58, https://doi.org/10.1177/1368430209356930. But for an alternative interpretation, see: Jacob M. Vigil and Chance Strenth, "Facial Expression Judgments Support a Socio-Relational Model, Rather Than a Negativity Bias Model of Political Psychology," *Behavioral and Brain Sciences* 37, no. 3 (June 2014): 331–32, https://doi.org/10.1017/S0140525X13002756.

176 **attracted by the negatively tinged information:** Luciana Carraro, Luigi Castelli, and Claudia Macchiella, "The Automatic Conservative: Ideology-Based Attentional

Asymmetries in the Processing of Valenced Information," *PLOS ONE* 6, no. 11 (November 9, 2011): e26456, https://doi.org/10.1371/journal.pone.0026456.

176 **tried to directly replicate:** Benjamin R. Knoll, Tyler J. O'Daniel, and Brian Cusato, "Physiological Responses and Political Behavior: Three Reproductions Using a Novel Dataset," *Research & Politics* 2, no. 4 (October 1, 2015): 2053168015621328, https://doi.org/10.1177/2053168015621328; Patrick Fournier, Stuart Soroka, and Lilach Nir, "Negativity Biases and Political Ideology: A Comparative Test across 17 Countries," *American Political Science Review* 114, no. 3 (August 2020): 775–91, https://doi.org/10.1017/S0003055420000131; Bert N. Bakker et al., "Conservatives and Liberals Have Similar Physiological Responses to Threats," *Nature Human Behaviour* 4, no. 6 (June 2020): 613–21, https://doi.org/10.1038/s41562-020-0823-z; Mathias Osmundsen et al., "The Psychophysiology of Political Ideology: Replications, Reanalyses, and Recommendations," *Journal of Politics* 84, no. 1 (January 2022): 50–66, https://doi.org/10.1086/714780; Kevin B. Smith and Clarisse Warren, "Physiology Predicts Ideology. Or Does It? The Current State of Political Psychophysiology Research," *Current Opinion in Behavioral Sciences*, Political Ideologies, 34 (August 1, 2020): 88–93, https://doi.org/10.1016/j.cobeha.2020.01.001.

176 **"disgust has been used throughout history":** Martha C. Nussbaum, *Hiding from Humanity: Disgust, Shame, and the Law* (Princeton University Press, 2009), 14.

177 **correlation between conservative policy preferences and disgust reactions:** Yoel Inbar, David A. Pizarro, and Paul Bloom, "Conservatives Are More Easily Disgusted Than Liberals," *Cognition and Emotion* 23, no. 4 (June 1, 2009): 714–25, https://doi.org/10.1080/02699930802110007; Kevin B. Smith et al., "Disgust Sensitivity and the Neurophysiology of Left-Right Political Orientations," *PLOS ONE* 6, no. 10 (October 19, 2011): e25552, https://doi.org/10.1371/journal.pone.0025552.

177 **study with thousands of Danish and American participants:** Lene Aarøe, Michael Bang Petersen, and Kevin Arceneaux, "The Behavioral Immune System Shapes Political Intuitions: Why and How Individual Differences in Disgust Sensitivity Underlie Opposition to Immigration," *American Political Science Review* 111, no. 2 (May 2017): 277–94, https://doi.org/10.1017/S0003055416000770.

178 **French word *sens*:** Noted in Maurice Merleau-Ponty, *The Phenomenology of Perception*, trans. Donald A. Landes (London: Routledge, 2012), 263. Quoted in Sara Ahmed, *Queer Phenomenology* (Durham, NC, and London: Duke University Press, 2006), Kindle, 133.

178 **"this world is there for me":** Edmund Husserl, *Ideas Pertaining to a Pure Phenomenology and to a Phenomenological Philosophy*, trans. F. Kersten (Dordrecht: Kluwer Academic, 1982), 53.

179 **ideology and pain sensitivity:** Spike W. S. Lee and Cecilia Ma, "Pain Sensitivity Predicts

Support for Moral and Political Views across the Aisle," *Journal of Personality and Social Psychology* 125, no. 6 (2023): 1239–64, https://doi.org/10.1037/pspa0000355.

179 **ideology and taste sensitivity:** Benjamin C. Ruisch et al., "A Matter of Taste: Gustatory Sensitivity Predicts Political Ideology," *Journal of Personality and Social Psychology* 121, no. 2 (2021): 394–409, https://doi.org/10.1037/pspp0000365.

179 **interoceptive sensitivity:** Benjamin C. Ruisch et al., "Sensitive Liberals and Unfeeling Conservatives? Interoceptive Sensitivity Predicts Political Liberalism," *Politics and the Life Sciences* 41, no. 2 (September 2022): 256–75, https://doi.org/10.1017/pls.2022.18.

180 **heightened physiological arousal:** Bert N. Bakker, Gijs Schumacher, and Matthijs Rooduijn, "Hot Politics? Affective Responses to Political Rhetoric," *American Political Science Review* 115, no. 1 (February 2021): 150–64, https://doi.org/10.1017/S0003055420000519.

180 **investigation by political psychologists in Amsterdam:** Maaike D. Homan, Gijs Schumacher, and Bert N. Bakker, "Facing Emotional Politicians: Do Emotional Displays of Politicians Evoke Mimicry and Emotional Contagion?," *Emotion* 23, no. 6 (2023): 1702–13, https://doi.org/10.1037/emo0001172.

181 **"Our emotional lives":** Angela Y. Davis, "Transnational Solidarities," speech delivered at Boğaziçi University, Istanbul, Turkey, on January 9, 2015, in *Freedom Is a Constant Struggle: Ferguson, Palestine, and the Foundations of a Movement*, ed. Frank Barat (Chicago: Haymarket Books, 2016).

181 **attentional and physiological responses to sights of inequality:** Attentional study: Hannah B. Waldfogel et al., "Ideology Selectively Shapes Attention to Inequality," *Proceedings of the National Academy of Sciences* 118, no. 14 (April 6, 2021): e2023985118, https://doi.org/10.1073/pnas.2023985118.

Physiological study: Shahrzad Goudarzi et al., "Economic System Justification Predicts Muted Emotional Responses to Inequality," *Nature Communications* 11, no. 1 (January 20, 2020): 383, https://doi.org/10.1038/s41467-019-14193-z.

CHAPTER 16: AN IDEOLOGY WALKS INTO A BRAIN SCANNER

184 ***Neural polarization***: Yuan Chang Leong et al., "Conservative and Liberal Attitudes Drive Polarized Neural Responses to Political Content," *Proceedings of the National Academy of Sciences* 117, no. 44 (November 3, 2020): 27731–39, https://doi.org/10.1073/pnas.2008530117; Daantje de Bruin et al., "Shared Neural Representations and Temporal Segmentation of Political Content Predict Ideological Similarity," *Science Advances* 9, no. 5 (February 2023): eabq5920, https://doi.org/10.1126/sciadv.abq5920; Noa Katabi et al., "Deeper Than You Think: Partisanship-Dependent Brain Responses in Early Sensory and Motor Brain Regions," *Journal of Neuroscience* 43, no. 6 (February 8, 2023): 1027–37, https://doi.org/10.1523/JNEUROSCI.0895-22.2022.

184 **studies employing machine-learning techniques:** Woo-Young Ahn et al., "Nonpolitical Images Evoke Neural Predictors of Political Ideology," *Current Biology* 24, no. 22 (November 2014): 2693–99, https://doi.org/10.1016/j.cub.2014.09.050.

186 **team of London-based researchers:** Ryota Kanai et al., "Political Orientations Are Correlated with Brain Structure in Young Adults," *Current Biology* 21, no. 8 (April 26, 2011): 677–80, https://doi.org/10.1016/j.cub.2011.03.017.

186 **independent research team in Amsterdam:** G. Schumacher, D. Petropoulos Petalas, and H. S. Scholte, "Are Political Orientations Correlated with Brain Structure? A Preregistered Replication of the Kanai et al. (2011) Study," unpublished research paper, n.d.

186 **study by researchers in New York:** H. Hannah Nam et al., "Amygdala Structure and the Tendency to Regard the Social System as Legitimate and Desirable," *Nature Human Behaviour* 2, no. 2 (February 2018): 133–38, https://doi.org/10.1038/s41562-017-0248-5.

187 **greater combined left and right amygdala volume:** Why did the New York team implicate the bilateral amygdalae while the London team implicated only the right amygdala? This may be due to the different functions that the left and right amygdala play in emotional processing, or perhaps it is because the researchers focused on slightly different ideological outcomes—conservatism versus system justification. The differences may also be due to the smallness of the amygdala, which makes it more difficult for neuroimaging devices to analyze relative to bigger structures and can therefore lead to uncertainty regarding whether only one of the amygdalae is involved or both; Jerry E. Murphy et al., "Left, Right, or Bilateral Amygdala Activation? How Effects of Smoothing and Motion Correction on Ultra-High Field, High-Resolution Functional Magnetic Resonance Imaging (fMRI) Data Alter Inferences," *Neuroscience Research* 150 (January 1, 2020): 51–59, https://doi.org/10.1016/j.neures.2019.01.009.

187 **amygdalae store emotional associations:** Dharshan Kumaran, Hans Ludwig Melo, and Emrah Duzel, "The Emergence and Representation of Knowledge about Social and Nonsocial Hierarchies," *Neuron* 76, no. 3 (November 2012): 653–66, https://doi.org/10.1016/j.neuron.2012.09.035.

187 **neural system named for its edgedness:** Luiz Pessoa and Patrick R. Hof, "From Paul Broca's Great Limbic Lobe to the Limbic System," *Journal of Comparative Neurology* 523, no. 17 (December 1, 2015): 2495–500, https://doi.org/10.1002/cne.23840.

188 **gradient of subdivisions:** R. J. Morecraft et al., "Cytoarchitecture and Cortical Connections of the Anterior Cingulate and Adjacent Somatomotor Fields in the Rhesus Monkey," *Brain Research Bulletin* 87, no. 4 (March 10, 2012): 457–97, https://doi.org/10.1016/j.brainresbull.2011.12.005.

188 **hub that has unusually high connectivity:** Wei Tang et al., "A Connectional Hub in the Rostral Anterior Cingulate Cortex Links Areas of Emotion and Cognitive Control," ed. David Badre and Michael J. Frank, *eLife* 8 (June 19, 2019): e43761, https://doi.org/10

.7554/eLife.43761; Suzanne N. Haber et al., "Prefrontal Connectomics: From Anatomy to Human Imaging," *Neuropsychopharmacology* 47, no. 1 (January 2022): 20–40, https://doi.org/10.1038/s41386-021-01156-6.

189 **connecting brain function to ideological thinking:** Some functional MRI studies implicating the ACC in reasoning about political issues: Drew Westen et al., "Neural Bases of Motivated Reasoning: An fMRI Study of Emotional Constraints on Partisan Political Judgment in the 2004 U.S. Presidential Election," *Journal of Cognitive Neuroscience* 18, no. 11 (November 1, 2006): 1947–58, https://doi.org/10.1162/jocn.2006.18.11.1947; Ingrid J. Haas, Melissa N. Baker, and Frank J. Gonzalez, "Political Uncertainty Moderates Neural Evaluation of Incongruent Policy Positions," *Philosophical Transactions of the Royal Society B: Biological Sciences* 376, no. 1822 (February 22, 2021): 20200138, https://doi.org/10.1098/rstb.2020.0138; Ingrid Johnsen Haas, Melissa N. Baker, and Frank J. Gonzalez, "Who Can Deviate from the Party Line? Political Ideology Moderates Evaluation of Incongruent Policy Positions in Insula and Anterior Cingulate Cortex," *Social Justice Research* 30, no. 4 (December 1, 2017): 355–80, https://doi.org/10.1007/s11211-017-0295-0.

190 **egalitarians were found to have greater ERN amplitudes:** Meghan Weissflog et al., "The Political (and Physiological) Divide: Political Orientation, Performance Monitoring, and the Anterior Cingulate Response," *Social Neuroscience* 8, no. 5 (September 1, 2013): 434–47, https://doi.org/10.1080/17470919.2013.833549; David M. Amodio et al., "Neurocognitive Correlates of Liberalism and Conservatism," *Nature Neuroscience* 10, no. 10 (October 2007): 1246–47, https://doi.org/10.1038/nn1979.

190 **Individuals with low religious zeal:** Generally, these patterns persist even after controlling for general conservatism, intelligence, and personality. See: Małgorzata Kossowska et al., "Anxiolytic Function of Fundamentalist Beliefs: Neurocognitive Evidence," *Personality and Individual Differences* 101 (October 1, 2016): 390–95, https://doi.org/10.1016/j.paid.2016.06.039; Michael Inzlicht et al., "Neural Markers of Religious Conviction," *Psychological Science* 20, no. 3 (March 1, 2009): 385–92, https://doi.org/10.1111/j.1467-9280.2009.02305.x. However, see divergence in results when using auditory Stroop task: Magdalena Senderecka et al., "Religious Fundamentalism Is Associated with Hyperactive Performance Monitoring: ERP Evidence from Correct and Erroneous Responses," *Biological Psychology* 140 (January 1, 2019): 96–107, https://doi.org/10.1016/j.biopsycho.2018.12.007.

190 **experiment with Mormon students in Utah:** Marie Good, Michael Inzlicht, and Michael J. Larson, "God Will Forgive: Reflecting on God's Love Decreases Neurophysiological Responses to Errors," *Social Cognitive and Affective Neuroscience* 10, no. 3 (March 1, 2015): 357–63, https://doi.org/10.1093/scan/nsu096; Michael Inzlicht and Alexa M. Tullett, "Reflecting on God: Religious Primes Can Reduce Neurophysiological Response to Errors," *Psychological Science* 21, no. 8 (August 1, 2010): 1184–90, https://doi.org/10.1177/0956797610375451.

191 **Patients with frontal lesions were more politically conservative:** H. Hannah Nam et al., "Toward a Neuropsychology of Political Orientation: Exploring Ideology in Patients with Frontal and Midbrain Lesions," *Philosophical Transactions of the Royal Society B: Biological Sciences* 376, no. 1822 (February 22, 2021): 20200137, https://doi.org/10.1098/rstb.2020.0137.

192 **Brains with lesions to the ventromedial prefrontal cortex:** Irene Cristofori et al., "The Neural Bases for Devaluing Radical Political Statements Revealed by Penetrating Traumatic Brain Injury," *Social Cognitive and Affective Neuroscience* 10, no. 8 (August 1, 2015): 1038–44, https://doi.org/10.1093/scan/nsu155.

192 **War veterans with injuries to the ventromedial prefrontal:** Wanting Zhong et al., "Biological and Cognitive Underpinnings of Religious Fundamentalism," *Neuropsychologia* 100 (June 1, 2017): 18–25, https://doi.org/10.1016/j.neuropsychologia.2017.04.009.

194 **Barcelona research team explored the neural signature:** Clara Pretus et al., "The Role of Political Devotion in Sharing Partisan Misinformation and Resistance to Fact-Checking.," *Journal of Experimental Psychology: General* 152, no. 11 (November 2023): 3116–34, https://doi.org/10.1037/xge0001436; Clara Pretus et al., "Neural and Behavioral Correlates of Sacred Values and Vulnerability to Violent Extremism," *Frontiers in Psychology* 9 (December 21, 2018), https://doi.org/10.3389/fpsyg.2018.02462; Clara Pretus et al., "Ventromedial and Dorsolateral Prefrontal Interactions Underlie Will to Fight and Die for a Cause," *Social Cognitive and Affective Neuroscience* 14, no. 6 (August 7, 2019): 569–77, https://doi.org/10.1093/scan/nsz034; Nafees Hamid et al., "Neuroimaging 'Will to Fight' for Sacred Values: An Empirical Case Study with Supporters of an Al Qaeda Associate," *Royal Society Open Science* 6, no. 6 (June 12, 2019): 181585, https://doi.org/10.1098/rsos.181585.

Also convergent findings by: Melanie Pincus et al., "The Conforming Brain and Deontological Resolve," *PLOS ONE* 9, no. 8 (August 29, 2014): e106061, https://doi.org/10.1371/journal.pone.0106061; Gregory S. Berns et al., "The Price of Your Soul: Neural Evidence for the Non-Utilitarian Representation of Sacred Values," *Philosophical Transactions of the Royal Society B: Biological Sciences* 367, no. 1589 (March 5, 2012): 754–62, https://doi.org/10.1098/rstb.2011.0262.

CHAPTER 17: SPIRALING IN AND OUT

203 **"The dislocation involved in switching from one passion":** Eric Hoffer, *The Passionate State of Mind and Other Aphorisms* (Titusville, NJ: Hopewell Publishers, 2006), 5.

204 **"It is astonishing what a different result":** George Eliot, *The Mill on the Floss* (New York: Bartleby, 2000), 96.

205 **Trier Social Stress Test:** Andrew P. Allen et al., "Biological and Psychological Markers

of Stress in Humans: Focus on the Trier Social Stress Test," *Neuroscience & Biobehavioral Reviews* 38 (January 1, 2014): 94–124, https://doi.org/10.1016/j.neubiorev.2013.11.005.

205 **people perform poorly on cognitive tasks:** Marion Fournier, Fabienne d'Arripe-Longueville, and Rémi Radel, "Effects of Psychosocial Stress on the Goal-Directed and Habit Memory Systems during Learning and Later Execution," *Psychoneuroendocrinology* 77 (March 1, 2017): 275–83, https://doi.org/10.1016/j.psyneuen.2016.12.008; Jessica K. Alexander et al., "Beta-Adrenergic Modulation of Cognitive Flexibility During Stress," *Journal of Cognitive Neuroscience* 19, no. 3 (March 1, 2007): 468–78, https://doi.org/10.1162/jocn.2007.19.3.468; Franziska Plessow et al., "Inflexibly Focused under Stress: Acute Psychosocial Stress Increases Shielding of Action Goals at the Expense of Reduced Cognitive Flexibility with Increasing Time Lag to the Stressor," *Journal of Cognitive Neuroscience* 23, no. 11 (November 1, 2011): 3218–27, https://doi.org/10.1162/jocn_a_00024; Grant S. Shields, Matthew A. Sazma, and Andrew P. Yonelinas, "The Effects of Acute Stress on Core Executive Functions: A Meta-Analysis and Comparison with Cortisol," *Neuroscience & Biobehavioral Reviews* 68 (September 1, 2016): 651–68, https://doi.org/10.1016/j.neubiorev.2016.06.038.

205 **Participants in the stress condition:** For stress-induction experiments and gender differences, see: Bart Hartogsveld et al., "Balancing Between Goal-Directed and Habitual Responding Following Acute Stress," *Experimental Psychology* 67, no. 2 (March 2020): 99–111, https://doi.org/10.1027/1618-3169/a000485; Elizabeth V. Goldfarb et al., "Stress and Cognitive Flexibility: Cortisol Increases Are Associated with Enhanced Updating but Impaired Switching," *Journal of Cognitive Neuroscience* 29, no. 1 (January 1, 2017): 14–24, https://doi.org/10.1162/jocn_a_01029; Vrinda Kalia et al., "Acute Stress Attenuates Cognitive Flexibility in Males Only: An fNIRS Examination," *Frontiers in Psychology* 9 (November 1, 2018), https://doi.org/10.3389/fpsyg.2018.02084; Sonia J. Lupien et al., "Effects of Stress throughout the Lifespan on the Brain, Behaviour and Cognition," *Nature Reviews Neuroscience* 10, no. 6 (June 2009): 434–45, https://doi.org/10.1038/nrn2639; Grant S. Shields et al., "Acute Stress Impairs Cognitive Flexibility in Men, Not Women," *Stress* 19, no. 5 (September 2, 2016): 542–46, https://doi.org/10.1080/10253890.2016.1192603.

206 **A small study with fifteen-month-old infants:** Sabine Seehagen et al., "Stress Impairs Cognitive Flexibility in Infants," *Proceedings of the National Academy of Sciences* 112, no. 41 (October 13, 2015): 12882–86, https://doi.org/10.1073/pnas.1508345112.

206 **less likely to suffer from an acute stressor:** A. Ross Otto et al., "Working-Memory Capacity Protects Model-Based Learning from Stress," *Proceedings of the National Academy of Sciences* 110, no. 52 (December 24, 2013): 20941–46, https://doi.org/10.1073/pnas.1312011110.

207 **natural experiment by comparing medical students:** J. M. Soares et al., "Stress-Induced Changes in Human Decision-Making Are Reversible," *Translational Psychiatry* 2, no. 7 (July 2012): e131, https://doi.org/10.1038/tp.2012.59.

CHAPTER 18: THE IMPORTANCE OF BEING NESTED

210 **"The truth is rarely pure":** Oscar Wilde, *The Importance of Being Earnest,* in *The Plays of Oscar Wilde* (Ware: Wordsworth Classics, 2000), 368.

211 **innovative study with men of Moroccan descent:** Clara Pretus et al., "Neural and Behavioral Correlates of Sacred Values and Vulnerability to Violent Extremism," *Frontiers in Psychology* 9 (December 21, 2018), https://doi.org/10.3389/fpsyg.2018.02462.

212 **social exclusion is one of the most powerful predictors:** For a set of recent experiments exploring the impact of social exclusion on different ideological worldviews, see: Andrew H. Hales and Kipling D. Williams, "Marginalized Individuals and Extremism: The Role of Ostracism in Openness to Extreme Groups," *Journal of Social Issues* 74, no. 1 (2018): 75–92, https://doi.org/10.1111/josi.12257; Michaela Pfundmair and Geoffrey Wetherell, "Ostracism Drives Group Moralization and Extreme Group Behavior," *Journal of Social Psychology* 159 (October 1, 2018): 1–13, https://doi.org/10.1080/00224545.2018.1512947; Emma A. Bäck et al., "The Quest for Significance: Attitude Adaption to a Radical Group Following Social Exclusion," *International Journal of Developmental Science* 12, nos. 1–2 (January 1, 2018): 25–36, https://doi.org/10.3233/DEV-170230; Holly M. Knapton, Hanna Bäck, and Emma A. Bäck, "The Social Activist: Conformity to the Ingroup Following Rejection as a Predictor of Political Participation," *Social Influence* 10, no. 2 (April 3, 2015): 97–108, https://doi.org/10.1080/15534510.2014.966856; Emma A. Renström, Hanna Bäck, and Holly M. Knapton, "Exploring a Pathway to Radicalization: The Effects of Social Exclusion and Rejection Sensitivity," *Group Processes & Intergroup Relations* 23, no. 8 (December 1, 2020): 1204–29, https://doi.org/10.1177/1368430220917215; Michaela Pfundmair, "Ostracism Promotes a Terroristic Mindset," *Behavioral Sciences of Terrorism and Political Aggression* 11, no. 2 (May 4, 2019): 134–48, https://doi.org/10.1080/19434472.2018.1443965; Michaela Pfundmair and Luisa A. M. Mahr, "How Group Processes Push Excluded People into a Radical Mindset: An Experimental Investigation," *Group Processes & Intergroup Relations* 26, no. 6 (September 1, 2023): 1289–309, https://doi.org/10.1177/13684302221107782; Jeffrey Treistman, "Social Exclusion and Political Violence: Multilevel Analysis of the Justification of Terrorism," *Studies in Conflict & Terrorism* 47, no. 7 (July 2, 2024): 701–24, https://doi.org/10.1080/1057610X.2021.2007244.

213 **researchers found that under conditions of scarcity:** Amy R. Krosch and David M. Amodio, "Scarcity Disrupts the Neural Encoding of Black Faces: A Socioperceptual Pathway to Discrimination," *Journal of Personality and Social Psychology* 117, no. 5 (2019): 859–75, https://doi.org/10.1037/pspa0000168; Michael M. Berkebile-Weinberg, Amy R. Krosch, and David M. Amodio, "Economic Scarcity Increases Racial Stereotyping in Beliefs and Face Representation," *Journal of Experimental Social Psychology* 102 (September 1, 2022): 104354, https://doi.org/10.1016/j.jesp.2022.104354; Amy R. Krosch and David M. Amodio, "Economic Scarcity Alters the Perception of Race,"

Proceedings of the National Academy of Sciences 111, no. 25 (June 24, 2014): 9079–84, https://doi.org/10.1073/pnas.1404448111.

214 **"The Other fixes me with his gaze":** Frantz Fanon, *Black Skin, White Masks* (London: Penguin Modern Classics, 2021), 89 and 95.

215 **"Of all things that move man":** Ernest Becker, *The Denial of Death* (New York: Free Press, 2007), Kindle, 11.

216 **simulated walk through a graveyard:** Luca Chittaro et al., "Mortality Salience in Virtual Reality Experiences and Its Effects on Users' Attitudes towards Risk," *International Journal of Human-Computer Studies* 101 (May 1, 2017): 10–22, https://doi.org/10.1016/j.ijhcs.2017.01.002.

217 **Iranian college students randomly assigned:** Tom Pyszczynski et al., "Mortality Salience, Martyrdom, and Military Might: The Great Satan Versus the Axis of Evil," *Personality and Social Psychology Bulletin* 32, no. 4 (April 1, 2006): 525–37, https://doi.org/10.1177/0146167205282157.

217 **Conservative American college students:** Pyszczynski et al., "Mortality Salience, Martyrdom, and Military Might: The Great Satan Versus the Axis of Evil."

217 **heightened concern for the environment:** Matthew Vess and Jamie Arndt, "The Nature of Death and the Death of Nature: The Impact of Mortality Salience on Environmental Concern," *Journal of Research in Personality* 42, no. 5 (October 1, 2008): 1376–80, https://doi.org/10.1016/j.jrp.2008.04.007.

217 **German environmentalists to be more authoritarian:** Markus Barth et al., "Closing Ranks: Ingroup Norm Conformity as a Subtle Response to Threatening Climate Change," *Group Processes & Intergroup Relations* 21, no. 3 (April 1, 2018): 497–512, https://doi.org/10.1177/1368430217733119; Immo Fritsche et al., "Existential Threat and Compliance with Pro-Environmental Norms," *Journal of Environmental Psychology* 30, no. 1 (March 1, 2010): 67–79, https://doi.org/10.1016/j.jenvp.2009.08.007.

217 **mortality salience also induces conservative shifts:** John T. Jost et al., "The Politics of Fear: Is There an Ideological Asymmetry in Existential Motivation?," *Social Cognition* 35, no. 4 (August 2017): 324–53, https://doi.org/10.1521/soco.2017.35.4.324; Armand Chatard, Gilad Hirschberger, and Tom Pyszczynski, "A Word of Caution about Many Labs 4: If You Fail to Follow Your Preregistered Plan, You May Fail to Find a Real Effect" (OSF, February 7, 2020), https://doi.org/10.31234/osf.io/ejubn; Brian L. Burke, Andy Martens, and Erik H. Faucher, "Two Decades of Terror Management Theory: A Meta-Analysis of Mortality Salience Research," *Personality and Social Psychology*

Review 14, no. 2 (May 1, 2010): 155–95, https://doi.org/10.1177/1088868309352321; Brian L. Burke, Spee Kosloff, and Mark J. Landau, "Death Goes to the Polls: A Meta-Analysis of Mortality Salience Effects on Political Attitudes," *Political Psychology* 34, no. 2 (2013): 183–200, https://doi.org/10.1111/pops.12005.

218 **depend on the specific nature of the threat:** Christopher M. Federico and Ariel Malka, "The Contingent, Contextual Nature of the Relationship Between Needs for Security and Certainty and Political Preferences: Evidence and Implications," *Political Psychology* 39, no. S1 (2018): 3–48, https://doi.org/10.1111/pops.12477.

218 **Terrorist threats often lead to shifts:** Amélie Godefroidt, "How Terrorism Does (and Does Not) Affect Citizens' Political Attitudes: A Meta-Analysis," *American Journal of Political Science* 67, no. 1 (2023): 22–38, https://doi.org/10.1111/ajps.12692.

218 **charisma of relevant leaders:** Spee Kosloff et al., "The Effects of Mortality Salience on Political Preferences: The Roles of Charisma and Political Orientation," *Journal of Experimental Social Psychology* 46, no. 1 (January 1, 2010): 139–45, https://doi.org/10.1016/j.jesp.2009.09.002.

218 **failures to replicate the mortality salience effects:** Alan J. Lambert et al., "Toward a Greater Understanding of the Emotional Dynamics of the Mortality Salience Manipulation: Revisiting the 'Affect-Free' Claim of Terror Management Research," *Journal of Personality and Social Psychology* 106, no. 5 (2014): 655–78, https://doi.org/10.1037/a0036353; Richard A Klein et al., "Many Labs 4: Failure to Replicate Mortality Salience Effect With and Without Original Author Involvement," *Collabra: Psychology* 8, no. 1 (April 29, 2022): 35271, https://doi.org/10.1525/collabra.35271.

218 **manipulations work especially well for people:** Clay Routledge and Jacob Juhl, "When Death Thoughts Lead to Death Fears: Mortality Salience Increases Death Anxiety for Individuals Who Lack Meaning in Life," *Cognition and Emotion* 24, no. 5 (August 1, 2010): 848–54, https://doi.org/10.1080/02699930902847144; Matthew Vess et al., "The Dynamics of Death and Meaning: The Effects of Death-Relevant Cognitions and Personal Need for Structure on Perceptions of Meaning in Life," *Journal of Personality and Social Psychology* 97, no. 4 (2009): 728–44, https://doi.org/10.1037/a0016417.

218 **reprioritize their values**: Philip J. Cozzolino et al., "Greed, Death, and Values: From Terror Management to Transcendence Management Theory," *Personality and Social Psychology Bulletin* 30, no. 3 (March 1, 2004): 278–92, https://doi.org/10.1177/0146167203260716; Kenneth E. Vail et al., "When Death Is Good for Life: Considering the Positive Trajectories of Terror Management," *Personality and Social Psychology Review* 16, no. 4 (November 1, 2012): 303–29, https://doi.org/10.1177/1088868312440046.

219 **Some cultures are better versed:** Christine Ma-Kellams and Jim Blascovich, "Enjoying Life in the Face of Death: East–West Differences in Responses to Mortality Salience," *Journal of Personality and Social Psychology* 103, no. 5 (2012): 773–86, https://doi.org/10.1037/a0029366.

219 **Islamist militants belonging to the Filipino jihadist group:** Arie W. Kruglanski et al., "What a Difference Two Years Make: Patterns of Radicalization in a Philippine Jail," *Dynamics of Asymmetric Conflict* 9, nos. 1–3 (September 1, 2016): 13–36, https://doi.org/10.1080/17467586.2016.1198042.

219 **militants in Sri Lanka and the Philippines:** David Webber et al., "The Road to Extremism: Field and Experimental Evidence That Significance Loss-Induced Need for Closure Fosters Radicalization," *Journal of Personality and Social Psychology* 114, no. 2 (2018): 270–85, https://doi.org/10.1037/pspi0000111.

219 **criminals convicted of ideologically driven crimes:** Katarzyna Jasko, Gary LaFree, and Arie Kruglanski, "Quest for Significance and Violent Extremism: The Case of Domestic Radicalization," *Political Psychology* 38, no. 5 (2017): 815–31, https://doi.org/10.1111/pops.12376.

219 **reflect on moments of humiliation:** Webber et al., "The Road to Extremism: Field and Experimental Evidence That Significance Loss-Induced Need for Closure Fosters Radicalization."

219 **reverse process of significance *gain*:** Katarzyna Jasko et al., "Rebel with a Cause: Personal Significance from Political Activism Predicts Willingness to Self-Sacrifice," *Journal of Social Issues* 75, no. 1 (2019): 314–49, https://doi.org/10.1111/josi.12307.

220 **study of over 4,000 school students:** Roberto M. Lobato et al., "Impact of Psychological and Structural Factors on Radicalization Processes: A Multilevel Analysis from the 3N Model," *Psychology of Violence* 13, no. 6 (2023): 479–87, https://doi.org/10.1037/vio0000484.

220 **students living in vulnerable environments in Spain:** Roberto M. Lobato et al., "The Role of Vulnerable Environments in Support for Homegrown Terrorism: Fieldwork Using the 3N Model," *Aggressive Behavior* 47, no. 1 (2021): 50–57, https://doi.org/10.1002/ab.21933.

220 **studies with adolescents in Sweden and adults in Indonesia:** Katarzyna Jasko et al., "Social Context Moderates the Effects of Quest for Significance on Violent Extremism," *Journal of Personality and Social Psychology* 118, no. 6 (2020): 1165–87, https://doi.org/10.1037/pspi0000198; Marta Miklikowska, Katarzyna Jasko, and Ales Kudrnac, "The Making of a Radical: The Role of Peer Harassment in Youth Political Radicalism," *Personality and Social Psychology Bulletin* 49, no. 3 (March 1, 2023): 477–92, https://doi.org/10.1177/01461672211070420.

221 **"A man need not be a Stalinist":** Czesław Miłosz, *The Captive Mind*, trans. Jane Zielonko (London: Penguin Classics, 2001), 30–31.

CHAPTER 19: OTHERWISE

225 **"There is a large measure of totalitarianism":** Eric Hoffer, *The Passionate State of Mind and Other Aphorisms* (Titusville, NJ: Hopewell Publishers, 2006), aphorism #28.

227 **"some exceptional cases of children":** Else Frenkel-Brunswik, "Intolerance of Ambiguity as an Emotional and Perceptual Personality Variable," *Journal of Personality* 18, no. 1 (1949): 132, https://doi.org/10.1111/j.1467-6494.1949.tb01236.x.

228 **Zadie Smith underscored this:** Zadie Smith, "Speaking in Tongues," *New York Review of Books*, February 26, 2009, https://www.nybooks.com/articles/2009/02/26/speaking-in-tongues-2/.

229 **"*otherwise* as in, a firm embrace":** Lola Olufemi, *Experiments in Imagining Otherwise* (London: Hajar Press, 2021), 7.

229 **"the artist cannot and must not take anything":** James Baldwin, "The Creative Process," in *Creative America* (New York: Ridge Press, 1962), https://openspaceofdemocracy.wordpress.com/wp-content/uploads/2017/01/baldwin-creative-process.pdf.

EPILOGUE

237 ***This hypothesis sounds very plausible*:** Phrase quoted from Hannah Arendt's interview with Roger Errera, October 1973.

237 ***possesses neither depth nor any demonic dimension*:** Phrase quoted from Hannah Arendt's letter to Gershom Scholem following the publication of *Eichmann in Jerusalem*.

INDEX

ABOUT THE AUTHOR

Dr. Leor Zmigrod is a prizewinning scientist and pioneer in the field of political neuroscience. She studied at the University of Cambridge as a Gates Scholar and has held visiting fellowships at Stanford, Harvard, and both the Berlin and Paris Institutes for Advanced Study. She was listed on *Forbes*'s 30 Under 30 in Science and has won numerous prizes, including the Women of the Future Science Award and the Glushko Prize. Her research has been featured widely in the media, including in the *New York Times*, the *Guardian*, *Financial Times*, and *New Scientist*.